To: Jim Woodward

You have helped make
the Day after Tomorrow
possible.

Sincerely,

Bill Ruse

THE WORLD BEYOND Tomorrow

WILLIAM RUSE
DONALD STANSLOSKI

TABLE OF CONTENTS

THE WORLD BEYOND TOMORROW

What will the world look like three decades from 2019? Today's millennials will have reached retirement age. The mid-point of the 21st century will be a year away. With a pair of Nostradamus-augmented reality glasses firmly in place, what will the world look like in the following ten areas?

1. Health Care
2. Transportation
3. Food
4. Housing
5. Retail
6. Workplace
7. Government
8. Finances
9. Climate
10. Wars

It's always interesting to look at the predictions of Nostradamus—and then let the pundits put their own spin on events. Nostradamus predicted a great war in 2018 between two or more countries that would change the world, plus a plethora of natural disasters. It isn't until 2025 that he predicts a world filled with peace and serenity.[1] In real time, 2018 featured many regional conflicts and more than its share of natural disasters. As this is being written, the great war predicted for 2018 has fortunately not yet materialized. Political historians might call the mid-term elections in the United States a great war between the Democrats and Republican parties.

Predictions of the future are both interesting and scary. The scary part can be attributed to jobs that fade into history. Consider the following predictions: [2]

Auto repair shops will disappear. A gasoline engine has 20,000 individual parts. An electrical motor has 20. Electric cars are sold with lifetime guarantees and are only repaired by dealers. It takes only 10 minutes to remove and replace an electric motor.

Gas stations will disappear. Parking meters will be replaced by meters that dispense electricity. Companies will install electrical recharging stations; in fact, they've already started. You can find them at select Dunkin Donuts locations.

Coal industries will disappear. Gasoline and oil companies will disappear. Drilling for oil will stop. So, say goodbye to OPEC!

Homes will produce and store more electrical energy during the day, which they will then use and sell back the excess to the grid. The grid will store and dispense it to industries that are high electricity users. Has anybody seen the Tesla roof?

A baby of today will only see personal cars in museums.

In 1998, Kodak had 170,000 employees and sold 85% of all photo paper worldwide. Within just a few years, its business model disappeared, and it went bankrupt. Who would have thought that would ever happen? What happened to Kodak will happen in a lot of industries in the next five to ten years and most people don't see it coming. Did you think in 1998 that, three years later, you would never take pictures on film again? With today's smart phones, who even has a camera these days except professional photographers? UBER is just a software tool—it doesn't own any cars—and it is now the biggest taxi company in the world!

Airbnb is now the biggest hotel company in the world, although it doesn't own any properties.

With **artificial intelligence, computers become exponentially better in understanding the world. In the USA, young lawyers already don't get jobs. Because of IBM's Watson, you can get legal advice (only the basic stuff so far) within seconds, with 90% accuracy, compared with 70% accuracy** when done by humans. So, if you study law, stop immediately. There will be 90% fewer lawyers in the future—what a thought!—and only omniscient specialists will remain. Watson already helps nurses diagnose cancer; it is 4 times more accurate than human nurses.

Facebook now has a pattern recognition software that can recognize faces better than humans. In 2030, computers will become more intelligent than humans.

Autonomous cars will become common: in 2018 the first self-driving cars are already here. In the next two years, the entire industry will start to be disrupted. You won't want to own a car anymore, as you will call a car with your phone, it will show up at your location, and it will drive you to your destination. You will not need to park it; you will only pay for the driven distance; and you will be able to be productive while driving. The very young children of today will never get a driver's license and will never own a car. This will change our cities because we will need 90-95% fewer cars. We will be able to transform former parking spaces into parks.

1.2 million people die each year in car accidents worldwide, including those caused by distracted or drunk driving. We now have one accident every 60,000 miles; with autonomous driving that will drop to one accident in six million miles. That will save more than a million lives worldwide each year.

Most traditional car companies will doubtless become bankrupt. Traditional car companies will try the evolutionary approach and just build a better car, while tech companies (Tesla, Apple, Google) will do the revolutionary approach and build a computer on wheels.

Insurance companies will have massive trouble because, without accidents, the costs of car ownership will become cheaper. Their traditional car insurance business model will disappear.

Real estate will change. If you can work while you commute, people will move farther away to live in a more beautiful or affordable neighborhood.

Electric cars will become mainstream by about 2030. Cities will be less noisy because all new cars will run on electricity. Cities will have much cleaner air as well. Electricity will become incredibly cheap and clean. Solar production has been on an exponential curve for 30 years, but you can now see the burgeoning impact, and it's just getting ramped up.

In health, the Tricorder X price will be announced this year. There are companies who will build a medical device (called the "Tricorder" from Star Trek) that will work with your phone and will takes your retina scan, your blood sample, and your breath into it. It will then analyze 54 biomarkers that will identify nearly any disease. There are dozens of phone apps out there right now for health purposes.

Many people have tried to predict the future and failed. Only some have succeeded. Part of this book will discuss predictability and end with some predictions on future health care cost challenges. However, the chapter begins with the disclaimer that predictions are often wrong. That said, here are some famous examples:

1. Harold Camping predicted the world would end in 2011. It didn't, but he had predicted other fateful dates and the world did not end on those dates either. He calculated his dates from material in the Bible.

2. The Flat Earth believers have been around for a long time, going back to the early Greeks (notably, Pythagoras) for example. By the time of Galileo, the Catholic Church agreed that the world was round, but punished him when he said that the Earth went around the Sun, which was contrary to the Church's belief.

3. Irving Fisher predicted in 1929 that "Stocks have reached what looks like a permanently high plateau." Three days later the market crashed.

4. Technology has spawned more wrong predictions than anything else recently.

   a. Ken Olson, founder of Digital Equipment Corporation, said he saw no reason anyone would want a computer in a home.

   b. Daryl Zanuck, a movie producer, said in 1946 that television would not last.

    c.   Clifford Stoll, an astronomer, predicted in 1995 that the Internet could not replace a newspaper or a classroom.

    d.   In 2007, Steve Ballmer, CEO of Microsoft, said the iPhone would never take an important part of the telephone market.

    e.   A Decca Records executive in 1962 said, "The Beatles have no future in show business."

    f.   Time Magazine predicted in 1966 that online shopping would never become popular because "women like to get out of the house, like to handle the merchandise, like to be able to change their minds."

5.   In 1873, Sir John Eric Erichsen, a British doctor appointed Surgeon Extraordinary to Queen Victoria, believed that the head, abdomen, and chest could not be operated on successfully.

Back to the present, which happens to be 2019. Excerpts from the previous article give a glimpse of the future from one writer's perspective. Will the fossil fuel companies follow in the footsteps of Kodak and Sears and be relics to be viewed in the Museum of Science and Industry in Chicago or one of the Smithsonian buildings in Washington, D.C.? Will Elon Musk's Boring Company let us travel by underground train from coast to coast non-stop in less time than a jumbo jet? And will we even need jumbo jets for domestic travel? See news about the Boring Company at:

In the following ten chapters, we will look at *The World Beyond Tomorrow* – if indeed a world remains.

About This Book

The World Beyond Tomorrow is a proverbial time machine. It will transport you to the year 2050—the mid-point of the 21st century. What will the world look like thirty years from now? The reader will find a series of ten short stories, each occupying a chapter, and each exploring a different scenario. Journey into the future to discover what your authors believe you'll find in 2050 in the following areas:

1.   Health Care

2. Transportation

3. Food

4. Housing

5. Retail

6. Workplace

7. Government

8. Finances

9. Climate

10. Wars

11. Religion

12. Changing Times

Each chapter attempts to cover several objectives. First, how are things today and how did we get to where we are in 2020? Next, how will technology available today affect the future? Then we include what other writers and futurists believe, in their own words, will happen in 2050. We fertilize this ground with our own predictions and ideas.

The World Beyond Tomorrow is meant to serve another purpose. It is an "idea" book. There are objects not yet invented that your authors believe could be possible with the quantum computers, augmented reality, and artificial intelligence available thirty years from now. We don't tell you what the objects are, because, for all we know, the ideas may already be on someone's drawing board – or CAD drawings for a 3D Printer!

Want a hint of the bottom line? We believe the mid-point of the 21st century will be the golden age of some of the scenarios we profile, including health care and transportation.

Finally, this book is meant to be apolitical. Although some of our predictions may cause concern for a specific political constituency today, we believe the needs of the future will coalesce around needed outcomes.

William Ruse

Donald Stansloski

Date: October 1, 2019

ABOUT THE AUTHORS

William (Bill) Ruse, BSPH, MBA, JD

Ruse's specialties are pharmacy, law, hospital management, teaching, writing and consulting.

1957 – Ruse started his career as a staff pharmacist and subsequently became chief pharmacist at Lima Memorial Hospital in Lima, Ohio.

1960 – Joined the staff at Blanchard Valley Hospital in Findlay, Ohio, where he served as chief pharmacist, director of personnel and purchasing, assistant administrator, and, subsequently, chief executive. In 1964, Bill served as president and CEO of the Blanchard Valley Health Association, a parent corporation of Blanchard Valley Hospital, founded in 1984.

Ruse is president emeritus of the Blanchard Valley Health Association, which is headquartered in Findlay, Ohio. Ruse retired from Blanchard Valley in 2001 after serving as its president and CEO for 36 years. The Association is one of the largest world health care delivery systems in Ohio, where it holds interests in hospitals, nursing homes, managed care, specialty health centers, and several subsidiary health-related corporations spread throughout Northwest and West Central Ohio. In recognition of his years of leadership, Blanchard Valley named its new 65,000 square-foot building on its Findlay hospital campus the "William E. Ruse Center."

Ruse's educational background and licensures include the following:

- Cottonwood High School – Cleveland, Ohio – graduated 1953
- Ohio Northern University – Ada, Ohio – BS in pharmacy, 1953 – 1957
- Xavier University – Cincinnati, Ohio – MBA in Hospital Administration, 1961 – 1963
- University of Toledo – Doctor of Jurisprudence – 1968 – 1972

- University of Findlay – Honorary Doctor of Healthcare Management – 1997
- Ohio Northern University – Honorary Doctor of Public Service – 2003
- Walden University – started PhD studies – fall – 2016
- Licensed Attorney – State of Ohio – 1972 to present
- Licensed Pharmacist – State of Ohio – 1957 – 2000 (allowed to expire)
- Life Fellow – American College of Healthcare Executives (LFACHE)

Donald W. Stansloski, RPh, Ph.D., Professor Emeritus, Ohio Northern University and Dean Emeritus, the University of Findlay

Dr. Stansloski has over 50 years of experience in pharmacy practice and education at all levels and in numerous venues. He has practiced hospital, community, and clinical pharmacy in five states and two countries, Mexico and Zimbabwe. He was a member of the faculty at the University of Nebraska. Later, he joined the faculty of Ohio Northern University College of pharmacy to institute the system of experiential education for the first time at the college. While at Ohio Northern, he served as a Fulbright scholar at the University of Zimbabwe, where he taught clinical pharmacy and clinical pharmacokinetics. He co-authored eight books on the proper use of drugs, mainly for the consumer market. He published several academic papers and presented hundreds of continuing education programs for physicians, pharmacists, and nurses. He advanced through all the academic chairs and served as Associate Dean before retiring.

After retiring, he continued to provide continuing education to health professionals around the country. He taught endocrinology over the Internet to Doctor of Pharmacy students in Ohio Northern University.

Nine years after leaving Ohio Northern University, he was offered the opportunity to found a new College of Pharmacy at the University of Findlay. After leading the college to full accreditation, he retired as the first emeritus dean in the 130-year history of the University of Findlay.

Stansloski's educational background and licenses include the following:

- Big Rapids High School, Big Rapids, Michigan. Graduated 1957
- Ferris State University, Big Rapids, Michigan BS pharmacy – 1961
- University of Nebraska, Lincoln, Nebraska MS – 1967
- The University of Nebraska, Lincoln, Nebraska PhD – 1970
- Licensed Pharmacist, State of Michigan 1961 to 2015 (ret.)

Other Books by Ruse and Stansloski

A Prescription for Healthcare Reform – Fact Book and Road Map

A Prescription for College and University Survival – Fact Book and Road Map

A Prescription for Hospital and LTC Survival – Fact Book and Road Map

Other Books by Ruse

The How-to Series – Technology for Seniors – Fact Book and Road Map

FORWARD

One major difference between humans and other species is our ability to imagine what the future will be like or what we would like it to become. And with that 'picture' in mind, we have an amazing creative ability to bring that future into being. All you must do to see evidence of these distinctly human traits is to compare the way animals lived 5000 years ago to today. Jungle animals still live in the jungle. But humans have transformed the way they live in the past few thousand years and especially the past 100 years.

The lives of Ruse and Stansloski illustrate this unique human trait. They have devoted their lives to envisioning the future and then pouring their heart and soul into making that future happen. In *The World Beyond Tomorrow*, the authors take us on a journey to 2050 illustrating how our world might unfold. With topics ranging from healthcare to housing and from finances to faith, it is a stimulating exercise to consider what might be.

Since no one on earth knows the future, about all we can say for sure is that there will be surprises that none of us foresee. The value of this book is not the perfection of its predictions but the stimulation of our imagination. Read, think, enjoy!

Gary Harpst

SPECIAL THANKS TO:

Gary Harpst

Rare is the author who does not receive inspiration from an event, colleague, or friend. Our first "Thank You" goes to Gary Harpst. During a presentation on future events, Gary talked about driverless cars in future years and the adverse impact they would have on existing industries and small businesses. That 90-minute speech inspired one of the authors of this book to look at the future in a more aggressive way – and *The World Beyond Tomorrow* began to take shape.

Mr. Harpst is a technological entrepreneur. He and two partners developed Solomon Accounting Software, one of the first comprehensive accounting software packages. The product was sold in world markets and eventually purchased by Microsoft.

Gary is presently President and CEO of Six Disciplines which works with purpose driven leaders to take their organizations to the next step of development and expansion.

Talk to Mr. Harpst and he will give credit for his many successes to his belief in our Lord. We asked him to be a guest author and write a chapter on *Religion Beyond Tomorrow* and he readily agreed to do so. Moreover, he has written the Forward to this book. Gary Harpst exudes a passion that comes from within and is shaped by his beliefs.

John Stanovich, RPh

Mr. Stanovich served as Assistant Dean in the College of Pharmacy at The University of Findlay in Findlay, Ohio, and as a professor at Ohio Northern University in Ada, Ohio. He is actively engaged in several entrepreneurial ventures.

John is a partner in Creative Solutions Enterprises, LLC, the three-partner group that co-authored this book. As a partner, John reviewed chapters of this book and made helpful suggestions for improvements and additions.

Thank you, John, for your assistance and support.

CHAPTER ONE

Health Beyond Tomorrow

Predicting the future of health care is like taming a cat. Even the present is clouded with shifting variables. Add in the differences in nearly 200 countries world-wide and Merlin's crystal ball is filled with more questions than answers. All the ten components of the World Beyond Tomorrow come into play in the discussion of the future of health care. This includes the final component, wars, as many of the advances of health over the decades can be traced to battlefield needs.

How does all we do impact on health care? Consider these vignettes:

1. Health Care – The starting point
2. Transportation – Impacts access and emergency care
3. Food – From malnutrition to obesity and its consequences
4. Housing – A pillar of life sustainability
5. Retail – From gyms to drugstores to tele-medicine
6. Workplace – Wellness-orientated or a sweatshop; payment plan determination
7. Government – Single payer system or private insurance driven
8. Finances – Are we getting what we're paying for? Determines cost. access and quality
9. Climate – Impacts disease and the environment
10. Wars – Creates an urgent need for change

If a time machine existed today, it would be interesting to go back thirty years and view health care at the beginning of the last decade of the 20th century. A good measurement is cost. In the United States, per capita expenditures for health care were $2,566.[3] In 2017, per capita costs had risen to $10,224. Table 1 reflects the cost of health care in the United States compared to other wealthy countries. [4]

Table 1 - Health consumption expenditures per capita, U.S. dollars, PPP adjusted, 2017

United States	$10,224
Switzerland	$8,009
Germany	$5,728
Sweden	$5,511
Austria	$5,440
Netherlands	$5,386
Comparable Country Average	$5,280
France	$4,902
Canada	$4,826
Belgium	$4,774
Japan	$4,717
Australia	$4,543
United Kingdom	$4,246

Notes: U.S. value obtained from National Health Expenditure data. Health consumption does not include investments in structures, equipment, or research.

The note at the bottom of the graph is interesting. What would the U.S. figure be if expenditures for structures, equipment and research were included? As an example, the U.S. spent over $171 billion on medical research and development in 2016.[5] Although it's difficult to break out medical equipment spending in the U.S., the top ten medical equipment suppliers in the world had $181.9 billion in revenue in 2017.[6] It's safe to speculate that the U.S. accounted for at least one-third of medical equipment purchases. When the cost of structures is entered into the cost of health care equation, costs escalate dramatically. The cost of construction for hospitals depends on where you live. Becker's Hospital Review lists costs in eleven major U.S. cities: construction costs range from a low of $285 to $455 in Las Vegas to a high of $475 to $760 per square foot in Honolulu.[7] Costs to construct a nursing home vary between $197 and $231 per square foot.[8]

Bury the preceding comments in a time capsule because the changes three decades from now will be dramatic. Barring a war or major universal disaster, the Ruse/Stansloski prediction for the future of health care, both in the U.S. and the world, is for stabilizing costs, longer lifespans, and cures or controls of today's most debilitating and life-threatening diseases. Major impacts on health care at the mid-point of the 21st century will come from:

- Quantum Computing
- Technological innovation including augmented reality
- DNA mapping and Genomics
- Pharmaceuticals
- Health System changes

Quantum Computing

How fast is fast? Faster than a speeding bullet? Faster than today's rockets? Faster than the latest and most powerful chip on your laptop? Ask this question to Google, NASA, and the Department of Energy and they will say it's 100 million times faster than your laptop. That computer exists today and is aptly named Road Runner. Superman, move over. You have a competitor in the speed race.

What does all of this have to do with health care? Quantum computing sets the stage for a revolutionary advancement in all the bullet points in the preceding paragraph. Speech recognition software will transcribe the spoken word more accurately than typing health data into a laptop. Technology will contain software that monitors your health continuously. DNA mapping will advance to the stage that diseases will be identified before they occur and eliminated at the point of identification. Pharmaceuticals will target, control, and eliminate chronic diseases. Health systems will feature hospitals where patients are sedated on admission and wakened when it's time to go home. Remember the movie, Coma? The pods are coming.

Say goodbye to health insurance companies as we know them. They will morph into a monitoring station where software analyzes a health care procedure, assigns a code, bills the payer (most probably government), and sends a credit to a provider's account. Co-pays will be automatically billed to the patient.

If all of this sounds a bit scary, change will take place in stages so we can get used to what's next. Quantum computing can take multiple pieces of information and process this information so we have a road map that will help break down a seemingly impenetrable wall. And not all will jump aboard the changed landscape. Take patients being treated in pods as an example. Many of us will still want to see, touch, and talk to our loved ones as they recuperate in a hospital. Choices will remain, but the fact is, we can overcome some of our worst health-related fears. The neurologist suggesting that tests show Lou Gehrig's disease will no longer be delivering a death sentence.

Changes brought about by quantum computing will be both favorable and unfavorable. Medical practitioners will no longer be burdened with more employees doing coding and billing than those treating patients. The other side of the coin is this: what happens to those with coding and billing skills? They will be the victims of advanced software capabilities.

Dr. Adrian Raudaschi has suggested several advancements in medicine and health care made possible by quantum computing. These include: [9]

- Radiotherapy: Radiation therapy is the most widely used form of treatment for cancers. Radiation beams are used to destroy cancerous cells or at least stop them multiplying. Multiple simulations can be run simultaneously.
- Drug Research: Larger sized molecules can be compared.
- Drug Interactions: Over 20,000 proteins encoded in the human genome can start to simulate their interactions with models of existing drugs or new drugs that haven't been invented yet.
- Artificial Intelligence: Quantum computing will allow doctors to compare much more data in parallel, simultaneously, and all permutations of that data, to discover the best patterns that describe it.
- Disease Screening: Quantum computing will enhance screening for diseases.
- Genomic Medicine: Using quantum computers, we can more quickly sequence DNA and solve other Big Data problems in health care. This opens up the possibility of personalized medicine based on an individual's unique genetic makeup.

While quantum computing cannot now compete with the complexity of the human brain, it can chart the building blocks of life. Once identified, diseases can be eliminated or modified, and life extended so that a 100-year-old person can still look forward to a decade of life.

As with most advances in the physical world, there is a dark side to quantum computing. Data privacy is a key concern. The hackers of the world will be countries, rather than individuals, simply because the cost of super computers will be beyond the financial capabilities of an individual.

Quantum computing can get us to the planets of our universe and beyond, but the technology can also start wars, shut down infrastructures, and in general create havoc beyond imagination. It can be more devastating than a nuclear war, as electricity, transportation, and communications can be shut down in an entire country. What is needed is a world body, initially composed of the world's leading superpowers. Note that we have bypassed the U.N. in this recommendation. Policies need to be adopted to preserve the advantages of the technology while completely curtailing its use for adversary purposes. Let's call this new body the QC World Congress (QCWC). The time to establish the QCWC is now, in the year 2020.

Impenetrable data spheres will surround data repositories. In homes, laptops and desktops will not be necessary. Computer rooms in the home will have a data sphere room. Inside the sphere will be a monitor screen. Talk to the screen and it will record your conversation or post a message on your friend's virtual Facebook account. A data repository sphere is noted below.

Technological Innovation

Hop aboard Captain Kirk's transponder and travel back to the 1990s. Our purpose will be to check out technology thirty years ago so we can get used to what will happen thirty years in the future from the year 2020. Smartphones will be a good starting point. The screenshot below shows a Siemens configuration for a family smartphone network.[10]

Back to the present and a family smartphone:[11] the 1990 system seems archaic by comparison.

The World Wide Web (WWW) was introduced in 1990, mostly as a tool for researchers. Today it is part of our daily lives.

The Apple Macintosh computer was introduced in 1990.[12] The retail version had one MB of memory.

Windows introduced Windows 3 that allowed PCs to support large graphical applications for the first time. A screenshot of Windows 3 appears below.

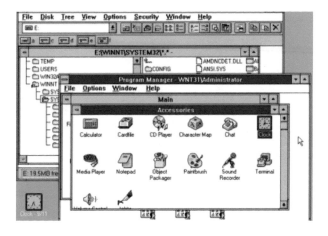

We are now up to Windows 10 and it runs most of the World's computers. Tablets, I-pods, Amazon, Kindle, Alexa and augmented reality have all appeared on the scene since 1990. Quantum computing has appeared although it is in its infancy.

Fast forward to the mid-point of the 21st century – 2050. Augmented reality, our AR, will lead, along with quantum computing, the technological revolution of the next thirty years.

AR is rapidly growing in popularity because it brings elements of the virtual world, into our real world, thus enhancing the things we see, hear, and feel. [13]

Plunk down $249 and pick up an Oculus AR headset on Amazon and start your visual journey. A screenshot of what you'll receive appears below.[14]

Oculus Go Standalone Virtual
Reality Headset - 64GB

Tele-Medicine is available today, perhaps as close as your local pharmacy. In 2050 you may have a doctor in your home. Turn on your TV, adjust your special glasses, tell your TV you want your doctor or other health care provider and, voila, you'll be connected to the health care provider of your choice. Of course, an appointment will be necessary, but that's easily done by communicating through your TV.

Need a blood test? You'll have a needle to prick your thumb and insert the blood sample into a home device that analyzes your blood and submits results electronically to your health care provider.

Home units will be developed to take basic x-rays in the home, interpret the results, and transmit them to your physician

Implanted chips are already available to measure many bodily functions. The Apple watch will take a single lead EKG. Project this technology forward thirty years, and the body becomes a tell-all snapshot of what's happening internally and externally, all by using a wearable or implantable chip.

Dr. Pam Miner weighs in with the following predictions for 2050: [15]

1. There has been a paradigm shift over the past 100 years, from taking action to eradicate the disease, condition, or thing we do not want to a more naturally-grounded assumption that the body contains

the ability to balance and heal itself, to contend with the stresses and invasions and to overcome them. In all likelihood, we will find additional environmental ways to help the body defeat illness and promote healthy wellbeing.

2. The growth of natural tissue and organs will not only be a treatment for organ failure—heart, lung or kidney—but will also impact spinal cord injuries, Parkinson's disease, and other neurological conditions. We will have improved implantable devices and more robotic procedures. Telemedicine will be used not only for consulting and treating but for integrating education with easily accessible and understandable health care information.

Health Care in the World Beyond Tomorrow will use AR to transform how and when health care is delivered. Personalized care will suffer, but outcomes will improve.

DNA Mapping and Genomics

Genomics is a branch of biotechnology concerned with applying the techniques of genetics and molecular biology to the genetic mapping and DNA sequencing of sets of genes or the complete genetic code of selected organisms, with organizing the results in databases, and with applications of the data. [16] The definition may be a mouthful, but the results are the key to disease identification and, once identified, disease elimination. Use of this technology can have social and political implications, especially at the embryonic stage of life. The guiding principle here is choice. Do I want the cancer or Alzheimer's gene to be eliminated from my unborn baby?

An easier question is whether to use genomics in the development of pharmaceuticals that will target and destroy a specific disease. More troublesome is the question of mental illness. Take substance abuse addiction as an example. If opiate addiction is a disease, as most believe it is, then DNA mapping and genomic medicine should be able to trace the gene and eliminate it. If there ever was a magic potion, this technology is the proverbial genie in Aladdin's Lamp.

Trent Green, a Senior Vice President of Legacy Health makes telling points about genomics in the next three paragraphs. [17]

Your genome will be sequenced at a cost equivalent to a cup of Starbucks coffee, and health providers will help you fine-tune a life plan for nutrition, fitness, and other life elements to stay healthy.

Cancers and conditions like Alzheimer's will be largely preventable through very early intervention based on genomic information and interconnected sharing of effective treatment results from cloud-based expert medical systems. Nanotechnology will monitor real-time vitals, and also serve in treatment roles—removing plaque in blood vessels, performing biopsies of tumors, and fighting infection at the site.

Your pharmacy can 3D print medications tailored to perform specifically for you in treating an ailment or maintaining your health. Advances in neuro-technology will address most of the mental health issues resulting from neuro-chemical and physical conditions and will give great insight into the behavioral aspects of mental illness. The health care "system" will mainly intervene only for trauma, births and neonates, and surgeries, including implanting replacement organs grown from your genomic print.

Mr. Green's comments about sequencing at the cost of a Starbucks cup of coffee is telling. Today you can grab a Starbucks coffee for less than $2.50. In 30 years at a 2.5% inflation rate, you'll spend $5.24—still a great deal to find out what's going on in your body. Of course, fine-tuning your life plan may include eliminating coffee. Sorry, Starbucks.

One of your authors was talking to his daughter about this book. She said, "Dad, I don't want to know what will happen thirty years from now. Anyway, I'll probably be dead." Put that thought on hold. If we can avoid war and a catastrophic natural disaster, the future of health care is bright. Also, we will ease into longevity; it will not spring forth overnight. For example, life expectancy in the United States is decreasing rather than increasing right now. Chronic diseases and addiction are the causes. While the rest of the world is living much longer (depending on country statistics), the U.S. is slipping a bit. There is, however, a pot of gold at the end of the rainbow. It's called DNA mapping and genomics. The gene manipulation for dealing with chronic pain is known due to a mutation in an Italian family. The addiction genome cannot be far behind. A 100+ lifespan is within our reach.

Pharmaceuticals

Today, when we think of drugs in the United States, we think of a supermarket type building with a pharmacy tucked as far back as possible. Yet, travel abroad and you'll find a pharmacy that sells mostly pharmaceuticals and perhaps a few beauty products. Their storefront advertising is usually a green cross such as the one depicted below.

The first known drugstore was opened by Arabian pharmacists in Baghdad in 754. [18] Some claim that the first drugstore in America was opened in New Orleans in 1823. They claim this is the first drugstore in America because the first registered pharmacist in America started it. The other first drugstore in America is located in Fredericksburg, Virginia, and dates to the time of the Revolutionary War. Mrs. George Washington was one of the patients of this drugstore, but it didn't have a registered pharmacist as its originator. [19] In the 1980s, pharmacies in the U.S. transcended from small mom and pop stores to the big box retail giants that sell everything from aspirin to grocery items and cleaning detergents.

In the next thirty years, pharmaceuticals and the stores in which they are sold will change dramatically. It is likely that the trend to include a prescriber in the same building as the pharmacy will continue and grow stronger. It would be a short step to add laboratory testing and X-Ray equipment as well. At that point, the place would look like a fully equipped health center for outpatient care.

Some pharmaceuticals may buck the trend, and traditional Chinese medicine is an example. It's been around since the second century BC and will likely be around at the middle of the 21st century. A recent issue of National

Geographic featured a pangolin, whose scales are used in Chinese medicine. [20] The scales are used to treat cancer, inflammation and other ailments. A photo of a pangolin appears below, followed by a photo of a Chinese chemist.

Discussing pharmaceuticals available in 2050 requires a review of two important issues. First, what will the drugs of the future do, and secondly, can we afford to pay for them? A significant change in pharmaceutical research has just been announced. Up until the present, research for cancer has been focused on finding drugs that treat advanced stages of the disease. The head of research at a major pharmaceutical company is directing research efforts on detecting and treating early stages of the disease.

As noted earlier in this chapter, a pharmacy can 3D print medications tailored to perform specifically for a patient in treating an ailment or maintaining their health. The combination of pharmaceutical research, quantum computing, and DNA mapping will result in a new class of drugs that target and eradicate disease.

People with an addiction to alcohol are familiar with the drug Antabuse. When on Antabuse, the user becomes violently ill if he or she drinks alcohol.

Drugs of the future will target the genes that crave an addictive substance and remove the craving. What will this mean to society? Eight to eighty-five percent of the prison population has some type of substance abuse issues. Crime and the prison population will be reduced substantially. Mental health workers will be able to turn their attention to other societal problems, including, but not limited to, poverty and the homeless.

Breakthroughs will occur in geriatric medicine as another new class of medications targets the natural effects of aging. This will not be the proverbial fountain of youth, but senior citizens will feel younger than their actual age. These drugs will be the Viagra of old age.

Drugs will target chronic disease before they become long-term medical conditions lasting for life. Obesity, diabetes, heart disease, stroke, asthma, and COPD will be eliminated with a combination of genomic pharmaceuticals, medical and surgical intervention.

Pharmacokinetics will play an important part in the future. This is the science of movement of drugs within the body including absorption, distribution, metabolism, and excretion. Every person has a different rate of handling these things for each drug used. Now we give patients an average dose as though we are all average for our pharmacokinetics. Dosing drugs individually will be of utmost importance because of the high potency and they will target specific pharmaceutical armamentarium.

Utopia doesn't exist without a cost. With a single drug today costing a record $2,000,000, can we afford to look at the World Beyond Tomorrow? If we ask an audience to raise hands if they believed drug costs in the future would be affordable, few hands would go up—perhaps none.

In 2012, the U.S. funded 44% the worlds biomedical research, down from 57% a decade earlier. The pharmaceutical industry figures are shown in the graph below. [21]

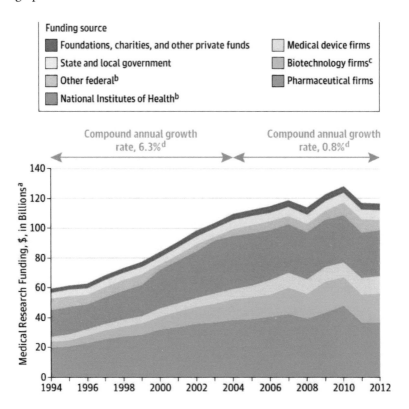

The mantra from the pharmaceutical companies is that the large amount spent on research justifies higher prices. In 2012 pharmaceutical companies spent $27 billion on advertising: nine out of ten big companies spent more on advertising than they did on research. [22]

By the mid-point of the 21st century, Pharmacy Benefit Managers (PBM) will be on a waning trail of existence. As the government will be the primary payer for pharmaceuticals, the middleman (PBMs) will not be needed. Delivery of drugs will adopt just-in-time delivery. Daily orders will be shipped by drones or air taxis directly from warehouses to pharmacies.

The government will negotiate drug prices that are similar to those paid by adjoining countries, Canada and Mexico. Pharmaceutical advertising will be virtually eliminated. Physicians will access TV for drugs and information via the web.

The U.S. government will increase the amount it spends on pharmaceutical research, but the U.S. will no longer subsidize research for the world.

With these changes, over $500 billion per year will be saved, measured in today's dollars. This will significantly mitigate the costs of pharmaceuticals in the future.

Health System Changes

Hospitals and health systems have proved they can adapt to change rapidly. The transition from a cost-based payment system to a prospective payment system is an example. In the latter, a fixed payment is received based on the medical diagnosis or procedure involved, regardless of the resources used. Virtually overnight, length of stays went down, and the number of outpatient procedures increased.

Thirty years ago, in 1990, there were 6,649 hospitals in the U.S. The number decreased to 5,534 in 2016 but spiked to 6,210 in 2017. [23]

A major impact on health systems will be the aging population. In 2020, 16.9% of U.S. citizens were 65 or older. By 2050, that figure is expected to grow to 22.1%. [24] The 5.2% increase translates to an additional 17 million senior citizens. Long term care facilities will be in demand, hospitals less so.

Those requiring hospital care will mostly require invasive procedures and surgery, including artificial organ transplants. Earlier, we mentioned pods where patients are sedated on admission, placed in a pod, fed intravenously, monitored continuously, transported to surgery, operated on, placed back in the pod and wakened after an appropriate recovery time or when therapy is needed. Since these scenarios will be difficult for the average (or even above average) person to understand, pod use will be limited to major teaching facilities.

Community hospitals will be dual function facilities, serving both as a hospital and as a long-term care facility (LTC). Rooms will be private, and wings will be capable of transitioning from acute to LTC interchangeably.

Most elderly citizens will remain in their homes, where they can comfortably age in place because of the changes we discussed earlier in this chapter. For those requiring nursing home care, the nursing home in 2050 will be markedly different than today's facilities. The authors described a future nursing home in their book, *A Prescription for Hospital and LTC Survival – Fact Book and Road Map*. Part of that discussion is repeated here. We'll call our hypothetical Nursing Home *Spring Lake*.

As your driverless car approaches Spring Lake, you'll notice no parking lot. Parking is all underground. You'll notice no handicapped parking spaces. As you pull into an empty parking space, sensors activate a transportation cart, which arrives automatically. You are transported to an elevator. The transportation cart returns to its parking space awaiting the next visitor or patient. When you return, facial recognition will allow the transportation cart to transport you to your car.

Reception will be your first stop. A hostess will welcome you, assign a room of your preference, and affix an electronic bracelet to your wrist. The bracelet will be pre-programed with all your medical and demographic information. A volunteer will arrive with a golf cart to take you on a tour to see the beauty shop, ice cream parlor, gift shop, and restaurant.

The electronic wrist band will serve as a telephone, which will be voice activated. Wrist bands will be provided to one or two family members so they can communicate with the patient. The patient's room number will be pre-programmed and provide directions to visitors. The wrist band will serve as a patient call button, putting the patient in touch with nursing personnel.

Rooms will be designated as suites and technology controlled beds will have mattresses that adjust to a patient's comfort level. Gone will be the wallboards with smiley faces and the names of staff members on duty. The TV will serve as both an entertainment center and an information center. Names of staff members, the day's scheduled patient activities, and menu selections will all be available. Specialties served in the restaurant, a smiley face progress chart, movies, and games will be available. The TV will be voice activated.

The floor will be pressure sensitive and can sense when a patient has fallen. Sensors will notify the nursing staff. A patient lift is stored in the ceiling and automatically descends when required.

Patients with visual problems will be issued electronic tablets that will contain all the information available on the in-room TV. Font sizes can be increased. A full range of games will be available, as well as movies.

The nursing home of the future features wide-open spaces outside the master suites. The restaurant replicates a fine dining establishment with a menu that changes daily. Patients can also order meals delivered to their room on an on-demand basis. Dietary restrictions will have been pre-programmed into their bracelets.

Major changes in health care as described in this chapter will help keep costs down. Moreover, long term care insurance will be mandatory by 2050, financed by employer and employee contributions.

The World Beyond Tomorrow in Health Care

The next thirty years will be known as the golden age of health care. Artificial body organ replacement will be commonplace. Think about what this can do for eyesight and other bodily functions that deteriorate with age. Chronic diseases can be identified in the unborn and eradicated. All the tools of chemistry and medicine will combine to provide a kaleidoscope of health inventions. There is a caveat here. Social systems must be capable of adjusting to an aging population that lives longer and consumes more resources. Read the next nine chapters.

CHAPTER TWO

Transportation Beyond Tomorrow

S enior citizens will recall the 1950s when newsstands and drugstore news racks carried a variety of pulp fiction magazines. For $0.25 you had your pick of dozens of magazines featuring subjects such as love stories, detective stories, and science fiction. A screenshot of a science fiction pulp appears below.

Pick up one of these pulp fiction books and you might read about what we can expect in the year 2050: everything from flying cars to laser guns. Today's millennials might get their future insights from the series of Star Wars movies.

Many of today's entrepreneurs in transportation from small businesses to major corporations will feel the pain of changes over the next thirty years. While they may not be extinct, their existence will be challenged. Troubled waters are ahead for:

- Fossil fuels industries, including petroleum companies and coal companies
- Quick change oil and lube shops
- Auto dealers
- Auto repair shops
- Auto parts manufacturers and suppliers
- Domestic airlines
- Diesel fuel powered trucks
- Highways will become less relevant
- State highway patrols will morph into state militias
- Stone companies and asphalt companies will diminish in size

Just as the mid-point of the 21st century will be the golden age of health care, the same is true for transportation. Consider the following vignettes.

- The vast majority of cars will be driverless and operated by electricity.
- City parking meters will become charging stations.
- Flying cars will be available—and replace most of domestic airline travel.
- Agricultural tractors will be capable of operating in the air, thus making planting impervious to weather conditions.
- High speed trains will operate underground—Elon Musk's Boring Company will be a stock market darling.
- Interplanetary transportation will become commonplace.
- Highways will be for trucks and the few human-driven cars still remaining. Speed limits will be automatically controlled,
- Drones will deliver packages securely to homes that have drop boxes installed.
- Few millennials (those born today will turn thirty in 2050) will own cars. They'll summon the Ubers of the world to take them from point A to point B.
- Parking lots will become pick-up stations.

For many, the projections listed above will be scary. Today we gather to watch classic car shows and vintage World War II planes displayed on Memorial Day. What will be displayed at the mid-point of the 21st century? Will fossil fuel museums become commonplace?

CEOs of companies affected adversely by advancements in technology must start to prepare their boards for dramatic change. To not do so is to replicate the Kodak and Sears implosions of the present. Three decades allows plenty of time to transition a business model, if the transition team is forward thinking. Take the coal industry as an example. For the most part, coal mines are located in beautiful surroundings—think Kentucky and West Virginia. Take advantage of this beautiful environment to create alternative business lines.

Let's explore more closely the transportation modes in 2050.

Driverless Electric Cars and Buses

The only thing holding back a significant increase in driverless cars in 2020 is local ordinances. True, the technology is in a test mode today, but progress will be rapid as the automotive industry pours billions of dollars in development. Congress must take control of driverless autonomous vehicle legislation. Note the screenshot below.[25]

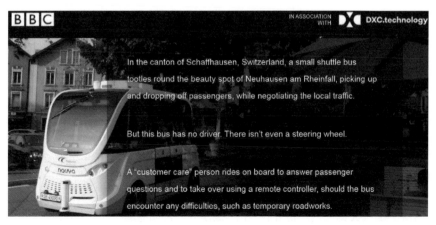

In the canton of Schaffhausen, Switzerland, a small shuttle bus tootles round the beauty spot of Neuhausen am Rheinfall, picking up and dropping off passengers, while negotiating the local traffic.

But this bus has no driver. There isn't even a steering wheel.

A "customer care" person rides on board to answer passenger questions and to take over using a remote controller, should the bus encounter any difficulties, such as temporary roadworks.

Imagine what we have today and put your imagination in a time machine and dial up the year 2050. Likely this bus will be driverless, electric, and run underground. The electricity will be generated by solar power that is stored and dispensed as needed.

Think of the advantage of driverless cars to people with disabilities. They will be able to use a car for transportation even if they are blind. The CDC estimates there are 53 million citizens with disabilities, and two-thirds of these have serious disabilities. A screenshot of the CDC data appears below. [26]

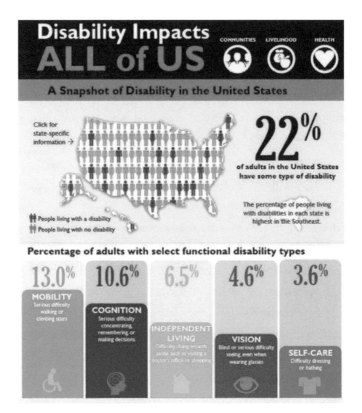

Cars will be driverless and airborne. Homes will have landing pads. Hop aboard and tell your car whether you want to take a highway or fly to your destination. Of course, you won't need to own a car. The Ubers of the world will supply all your transportation needs. A photo of a flying car appears below.

What impact will driverless cars have on the world? Consider the following: [27]

1. We'll be less likely to own a car. Why buy when you can summon a car on a moment's notice?
2. Urban centers will evolve. Streets can be narrower and pedestrian friendly. Traffic lights could become obsolete.
3. Business will come to us. The driverless truck may be a food store that comes to you.
4. More people can live independently. Illness, old age or a disability will not cause a transportation problem.
5. Donor organs could be in short supply. Why? Fewer accidents. But organs will be grown in a lab.
6. People will re-think their living arrangements. Long commutes will become acceptable.

Considering there are approximately 2,800 traffic related fatalities per month in the U.S., driverless vehicles could reduce these accidents by 50%.

To get around a big city, a solar powered driverless mini- scooter will transport you simply hop aboard and give your destination. Lanes will be devoted to mobile mini scooters, just as they are for bicycles today, Inclement weather?

No problem. A voice activate force shield will surround the passenger, keeping him or her cool in summer and warm in winter—and dry in all seasons.

Parking Meters Will Become Charging Stations, But Parking Lots Will Become Mostly Obsolete

Parking meters are both a curse and a blessing. Unbeknown to most, they can be a recruiting tool. One of your authors lives in a small city in the Midwest. He was trying to recruit a physician who lived in New York City. After a tour of the community, they were walking through the downtown area on their way to lunch. The physician saw the parking meters and ask how much it cost to park and what was the fine if the time elapsed. In those days, the hourly rate was a nickel and the fine a quarter. The doctor stopped dead in his tracks. He reiterated the New York fees, and if you exceeded your time limit—as he had on several occasions—his car would be towed, and it cost him a day off work to reclaim his vehicle. The doctor said, "Find me an office; I'm coming to your town." Sometimes it's the little things that count.

To the extent parking meters are still present in 2050, they will be charging stations. Actually, parking meters as charging stations have been around since the early part of the 21st century. [28]The newest EV (Electric Vehicle) charging stations in Austria were officially activated in the town of Ybbs an der Donau (colloquially referred to as "Ybbs") over the weekend. The new stations, like the rest of Ybbs and several of the surrounding neighborhoods, are powered by Wüster power stations, which claim that 100% of electricity comes from renewable sources. Regardless of the source of power, its distribution is rather clever. Using a system designed by Siemens, Ybbs residents can now recharge EVs from parking meters. A photo of the parking meter charging station appears below.

This Photo by Unknown Author is licensed under CC BY-SA-NC

Thirty years from now, parking meters will have disappeared but charging stations will take their place. These meters will be sparingly used as most vehicles will run on solar power. EVs (electric vehicles) will become SVs (solar vehicles). A prototype of a solar vehicle appears below.

Parking lots will become obsolete and the few that remain will be located underground. Forget about driving around and looking for a parking space. Place your car on an elevator and walk away. The car will be automatically parked and brought to you wherever you are located when it's time to leave. Settle back, select your favorite libation, and click on your on-board TV to watch the news or a movie.

Today's veterans remember Ducks—vehicles that can run on land and water. A photo of a World War II and Korean War Duck appears below.

In 2050, cars will be multi-dexterous. They will be at home on the road, in the air, or on the water. Take your car on a cruise down the Danube River in Europe or the Mississippi River in the United States. The mid-point of the 21st century will truly be the golden age of transportation.

Agricultural Transportation in The World Beyond Tomorrow

The giants of the agricultural revolution in the 18th century and early part of the 19th century included Eli Whitney (the cotton gin), Cyrus McCormick (the mechanical reaper), and John Froelich and John Deere (the mechanical tractor). All these inventions were part of the industrial revolution.

Today, farmers can watch their favorite television show as they plant or harvest the crops that feed much of the World's population. The amount of production per acre depends on many variables, not the least of which is Mother Nature. As this book is being written, many farmers in the Midwest have yet to plant a single seed through the middle of June. Torrential rains and floods have brought the farming community to its knees. In some parts of the country, planning is most delayed in recorded history. The screenshot below, published by National Geographic, shows rain drenched fields near Gardner, Illinois. [29]

It is probable that an entire year of agricultural production will be wiped out in the Midwest. Pundits have renamed the Weather Channel the "Water Channel." Farmers may decide to plant rice.

Help is on the way. By the mid-point of the 21st century, crop planters will hover above the ground and laser plant seeds in the soil below, regardless of weather conditions. Standing water pools will be sucked into a giant vacuum and deposited in a retention pond.

If the weather interferes with a harvest, crops will be harvested at ground height and technologically dried in drying bins. Tomorrow's tractors and tillers will allow fields to be plowed in the winter, even in sub-freezing temperatures.

Fertilizers in use in the 2050s will be environmentally friendly. All crops will be considered organically grown.

Regardless of the advancements in agricultural technology, it is probable there will be sporadic food shortages throughout the world. The shortages will be traced to population growth. The current world population of 7.6 billion is expected to reach 8.6 billion in 2030, 9.8 billion in 2050 and 11.2 billion in 2100. With roughly 83 million people being added to the world's population every year, the upward trend in population size is expected to continue, even assuming that fertility levels will continue to decline. [30] If the projections we have set forth in the chapter on Health Care Beyond Tomorrow hold true, the U.N.'s projections are on the low side. While produce may be grown in the home, it will take significant advances in agricultural technology to feed a hungry world. Biospheres on the moon are a reasonable possibility.

High Speed Trains in The World Beyond Tomorrow

In the U.S., high speed trains reach a top speed of 120 miles per hour, with an average speed of 80 mph. In Taiwan and Italy, high speed trains can travel 186 miles per hour. In China, the top speed reaches 220 mph. Why the difference? Politics. Local jurisdictions and some state regulations stand in the way of progress—and will continue to do so. Moreover, the U.S. does not have the technology in place, nor the manufacturing capabilities, to produce high speed trains. Japan has both and China does also. China has the longest high-speed train network in the world. The chart below shows China's rail network in kilometers (km). In 2020 it is projected that 30,000 km of rail will exist. This equates to 18,000 miles. [31]

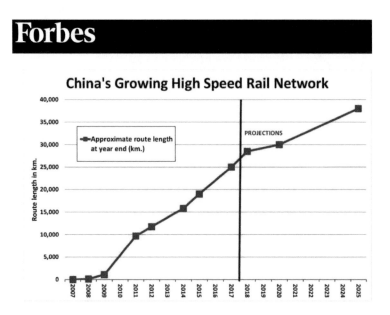

Dig a little deeper and see the future of high-speed train service in the U.S. The Boring Company, an Elon Musk enterprise (he of Telsa fame), plans to create tunnels underground that will provide railways—and car highways—for underground travel. Passengers will be able to travel from New York to Los Angles faster than taking a supersonic jetliner.

Today, The Boring Company provides pre-built tunnels that can be bored underground. Moreover, a pricing calculator will give an immediate cost estimate on a project. The screenshot below shows the type of tunnels available in 2019. [32]

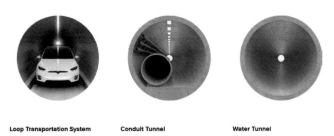

Estimated project pricing can typically be provided within 1 week. Stay tuned for the Tunnel Price Calculator coming to this site in 2019, where the user can enter product line, location, geology type, and length, and the calculator will return a project price range maximum and minimum.

Loop Transportation System   **Conduit Tunnel**   **Water Tunnel**

By the year 2050, The Boring Company's approach to using underground tunnels will allow cars and high-speed trains to travel coast to coast and points in between. It will be the subway system of the future. Moreover, high speed internet fiber cables can share transportation tunnels. A grim use will be the use of the tunnel network as bomb shelters. As the reader will note in a subsequent chapter of this book, the authors do not expect the next war to be fought with bullets or bombs.

Interplanetary Travel in the World Beyond Tomorrow

Man has set foot on the moon after a 239,000-mile journey. The International Space Station hovers 220 miles above the earth. Space probes have reached the far end of our galaxy. The U.S. has targeted Mars as the next planet for human landing—and Mars is 231 million miles away. Virgin Galactic is taking reservations for a visit to outer space. Blue Origin has this to say about living and working in space: "Blue's vision is a future where millions of people are living and working in space. In order to preserve Earth, our home, for our grandchildren's grandchildren, we must go to space to tap its unlimited resources and energy. If we can lower the cost of access to space with reusable launch vehicles, we can all enable this dynamic future for humanity." Guess who owns Blue Origin? None other than Jeff Bezos, founder, chairman, and CEO of Amazon.

With the combined technology savvy of Virgin Galactic, Blue Origin, and The National Aeronautics and Space Agency (NASA), we can expect the moon to be populated with humans by 2050, and there will be more space stations where humans will work, play, grow their own food, and produce water. The moon will afford mining as well as living capabilities where domed cites flourish. The Amazing Magazine stories of the 1950s will make good reading and allow us to visualize the future.

China and Russia will compete with the United States for space superiority, just as Russia did during the cold war of the 1960s. It will take a significant effort by the U.S. government to keep this country ahead in technology. The task will not be easy. As an example, in the city of Chennai, India, they graduate more engineers than the entire U.S. It will take a herculean effort by the United States to maintain its lead in outer space development and exploration. President Kennedy's words come to mind: "By the end of this decade, we will send men to the moon and return them safely to earth".

A screenshot of Jeff Bezos's lunar landing spaceship appears below. [33]

CHAPTER THREE

Food Beyond Tomorrow

L ife has two basic necessities – food and water. Depending on where you live, either can be scarce or limited because of the income of an individual or family unit. In the United States, no one should go without access to food of some type, yet in the rest of the world there are many regions where malnutrition and starvation are the norm. As we pointed out in Chapter Two, the current world population of 7.6 billion is expected to reach 8.6 billion in 2020, 9.8 billion in 2050, and 11.1 billion in 2100. The world's population is growing by nearly 83 million per year. Moreover, the advances in health care that we profiled in Chapter One will mean people live much longer, thus increasing the population exponentially.

Climate change is bringing an increased number of floods and draughts. As this is being written, the Midwest in the U.S. has experienced record flooding. In June 2019 many millions of acres of farmland are still too wet for planting and more rain is forecast for the weeks ahead. For the U.S, these weather conditions have a two-pronged effect—higher prices for food at home and less income for farmers, both from domestic production and crop exports.

A recent article by Joseph Hincks in Time Magazine summed up the world food crisis. The world currently produces more than enough food to feed everyone, yet 815 million people (roughly 11% of the global population) are starving in 2016, according to the U.N. By 2050, our food supplies will be under far greater stress. Demand will be more than it is today, but climate change, urbanization, and soil degradation will have shrunk the availability of arable land, according to the World Economic Forum. Add water shortages, pollution, and worsening inequality into the mix and the implications are stark.[34]

Consider the adverse food impacts Mr. Hincks points out: [35]

- Climate change
- Urbanization
- Soil degradation
- Water shortages
- Pollution
- Worsening inequality

Before those Americans that can afford to do so start hording food, let's investigate the 2050 crystal ball to see what the future holds.

Biotechnology, agricultural innovation, and interplanetary colonization will help the food supply keep up with population growth. Organic fertilizer will be available by 2050. There will be less stress on the environment as driverless cars and buses and underground trains, fueled by solar powered cells reduce pollution and lead to a cleaner environment.

All of these factors combined lead to the realization that we can grow more per tillable acre, perhaps 25 to 70% more, thus keeping up with the increased population growth. [36]

In Chapter Two, we discussed flying tractors that can hover above the soil and laser-plant seeds in inclement weather. Also, in Chapter Two we discussed colonization of the moon, where, in domed cities, crops can be grown.

Genetic engineering offers a time-saving method for producing larger, higher-quality crops with less effort and expense.[37] An image of an insect resistant tomato plant appears below. [38]

Creation of an Insect Resistant Tomato Plant

Bacterium

DNA →

Insect resistance gene

1. Cut out the gene.

2. Insert gene into a vector with a selectable antibiotic resistance marker gene.

Antibiotic resistance gene

3. Copy vector in bacteria.

4. Coat tungsten or gold particles with DNA vectors.

5. Load vector-coated particles onto teflon bullet.

6. Load bullet into gene gun.

Gene Gun

7. Shooting the gene gun releases the particles at a high velocity penetrating the plant cells.

8. The vector enters the cell. The genes are incorporated into the plant genome.

9. The cells are plated on a selective antibiotic media. Only cells that have incorporated the vector will grow.

10. These cells are transferred to medium containing plant growth factors.

Insect resistant tomato plant

Genetics and farming are no strangers. Corn yields per acre have increased six-fold since the 1940s. In the U.S., corn yield exceeds 163 bushels per acre while the rest of the world averages 87 bushels per acre. [39] Although these figures are somewhat dated, farmers in the United States have used the advantages of the biotechnology science of genetics and better fertilizers to dramatically increase yields. In 2018, U.S. corn yields exceeded 176 bushels per acre. Think ahead to 2050 when mega-farms will populate the agricultural community. Domed farms will make farming impervious to weather.

Meat will give way to meatless concoctions made from soy and other plant foods. The trend is already under way. The Impossible meatless burger, taco, pizza, and baos (a Cantonese street food) is here now. The Impossible burger (White Castle serves them as sliders) contains soy protein, potato protein, coconut oil, sunflower oil, and thyme. Open up the Impossible.com web site and you will find the firm's mission statement: "To Save Meat. And Earth." A screenshot of the Impossible web site appears below. [40]

By 2050, meat consumption will fall by 50% and meatless products will the food of choice to replace meat. There will also be a combination of meat/non-met products much like the ethanol/gasoline combination today. For grilling out on Father's Day, order a bag of meatless ingredients, add water, mold into a patty, and grill.

Highrise buildings will become farmhouses where residents can grow their own food. A photo of the concept appears below. [41]

Interplanetary travel will result in new food products for use aboard spaceships and foods suited to the environment in which they are produced.

The food chain will change in the years ahead. There will be enough food to feed an increased world population, but access will continue to be a problem, as well as individual financial resources. Food quality and nutritional value will improve, and obesity will decrease in the developed nations of the world.

CHAPTER FOUR
Housing Beyond Tomorrow

D epending on where you live in the world, the type of housing you live in is often dictated by the environment, especially in poorer nations. Some Eskimos still live in igloos in Canada's Central Arctic region and Greenland. In the Sinai, some Bedouin people still live in tents. Many homeless people in the U.S. will pitch a tent under a bridge or other covered structure. A photo of a Bedouin family appears below. [42]

Next to water and food, shelter ranks high on the life sustainability index. The U.S. has a mix of housing ranging from crowded tenements in large cities to mega-million-dollar mansions in the suburbs to ranches spread over 1,000 square miles. Add farmland with farmhouses and barns and neighborhoods filled with condos to skyscrapers that reach the clouds, and you have housing to suit any taste or pocketbook.

The Veterans' Emergency **Housing** Program started **after World War II** was the most significant turning point in the history of North American cities.

A massive **housing boom following World War II** focused on car-oriented suburban development, [43] These often cookie-cutter homes dotted the suburban landscape—and they were affordable. These one-story ranch type homes were not beautiful, but rather functional. The tragedies of war often beget an increase in living standards once the wars have ended. The stately Victorian homes in the central city, and often were two-family dwellings.

The typical suburban home today will have a family room, kitchen, living room, dining room, and one to three bedrooms. At least 1½ baths is normal, as are a two-car garage and a laundry room. A basement and upstairs may also be present. Luxury homes have a library, exercise room, and an office. Add a home theater to really make the neighbors envious.

In the last thirty years housing has changed little except for the consequences of the 2009 recession when homeowners found their homes were worth less than the mortgage they owed. More houses can be found at the opposite end of the spectrum—bigger houses on one hand and small or tiny houses on the other hand. A photo of a small 1,000 sq. ft. that retails for less than $47,000 appears below. Shipping is free. [44] Tiny houses can be purchased for less than $5,000.

A tiny home available from Amazon and measuring 113 sq. ft. sells for $4990. The cabin appears below.

Purchasing your house is only part of the equation. You'll need a plot of land and the patience to construct a home after all the parts arrive. Years ago, when Sears was the Amazon of today, you could order a steel house from the Sears catalogue. Many of those homes are still in existence. Four hundred styles were offered. Today firms such as Schumacher will build a preassembled home on your lot.

A clothing company in Denmark is planning on building a skyscraper in a rural area. A screenshot of the building appears below. [45]

Homes in 2050

To visualize the home of the future, start with the present and think "smart." Everything from the full-wall TV to the timed doggy food station will be voice activated. R2D2, he of *Star Wars* fame, may be available to help with daily chores. Garages will be less important and replaced by landing pads for your flying car. A voice activated dome will cover your car when not in use.

A full wall-sized TV will serve dual functions as an entertainment center, shopping mall, computer, and tele-medicine device. The TV screen will be encased in a material that is impervious to quantum computing hackers.

The interior of houses will be more modular, changing to suit needs during the course of the day and over your lifetime. Walls on rollers will allow you to reconfigure your space from office, to lounge, to bedroom. [46]

Older homes will be capable of being retrofitted to take advantage of the new technological innovations. New homes will be heated and cooled with solar panels. A sketch of a family home in 2050 appears below. [47]

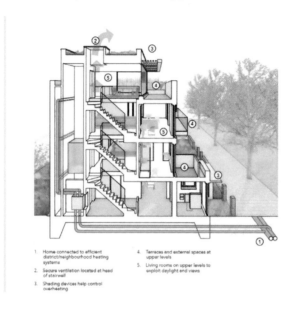

1. Home connected to efficient district/neighbourhood heating systems
2. Secure ventilation located at head of stairwell
3. Shading devices help control overheating
4. Terraces and external spaces at upper levels
5. Living rooms on upper levels to exploit daylight and views

Appliances will all be voice activated and, when need be, communicate with one another. At 7 AM the coffee maker may communicate to the kitchen thermostat to raise the temperature to 70 degrees Fahrenheit.

The smart floor will know when you fall and ask if help should be summoned. If you fall, place the palm of your hand on the floor and the floor will recognize you. Of course, if your fall is due to having too many martinis, you may tell the floor to shut up. It will say, "Yes, sir" or "Yes, miss."

Since 71% of the Earth's surface is covered with water, the boats and ships will have competition—from floating cities. The cities will be built on floating concrete platforms, each about five acres and holding 300 residents. A screen-shot of the city appears below. [48]

The most ambitious housing project of the future come from the fertile mind of Jeff Bezos, he of Amazon fame. Bezos has laid out a vision of having up to a trillion humans living in manufactured worlds, to be built by future generations. A photo of Mr. Bezos's vision appears below. [49]

Bezos envisions millions of colonies built on the moon, asteroids and other parts of the solar system. A photo of a colony appears below. [50]

Technology in the Home Beyond Tomorrow will be something one would expect to find in a science fiction magazine. Consider the following: [51]

- Smart refrigerators will give you cooking directions after you select the food you want.

- A scanner scans your hung clothes, finds the dirty spots and cleans them so you always have fresh, clean clothes in your closet.

- You can print your own clothes in your home. See image below.

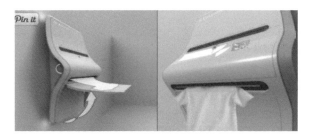

- Imagine your food, and you shall have it. Robots see your desired food recipes and get it ready for you without you having to place an actual spoken order. See image below.

Tomorrow's Home – A Blessing for Seniors

Seniors and their children will find much to love in the technologically advanced homes of the 2050s. Consider the following vignettes about a hypothetical widow, Madeline—we'll call her Maddie—and her cat, Fluffy.

- Maddie is suffering from moderate dementia. If left alone, she'll wander outside her home, often in her nightgown.

- Fluffy's litter box often goes unintended and Maddie forgets to feed her from time to time.

- Maddie will forget to take her medication or eat regular meals.

- Maddie's daughter, Christine, finally decides to place her mother in the dementia unit of a long-term care facility.
- Maddie's moderate assets dissipate rapidly as the LTC facilities monthly bills are paid.

Fast Forward to 2050. Let's check in on Maddie who suffers from moderate dementia.

- Maddie is living at home. Christine keeps tabs on mom via TV and her mobile phone.
- Doors to the home remain locked. Special alarms in the house will alert the fire department if a fire breaks out, and the outside doors will open automatically.
- Christine has gathered facial recognition information of Maddie's friends, helpers, and minister, and the door will unlock if they visit.
- Christine can tell if her mom has forgotten to take her medication or eat a meal.
- Maddie's care team checks her physiological health through implanted chips in Maddie's arm. Visits are made via TV in Maddie's home.
- If she falls, sensors in the floor will alert her daughter and emergency personnel.
- If there are lifesaving technologies Maddie can't afford, the federal government's national health plan will pay the purchase and installation price. The investment is less expensive than paying LTC fees.
- Fluffy is happy. His human is at home!

The home of tomorrow will be a cave for all seasons. Come home from work, plop down on your theater seat, tell the TV to give you a news update, put on your headset, and order your meal and a beer, scotch, or glass of wine and think to yourself, it doesn't get much better than this.

CHAPTER FIVE

Retail Beyond Tomorrow

Less than 30 years ago, in 1994, a 30-year-old young man decided selling books over the internet might be profitable. The young man was Jeff Bezos and the company he started out of his garage was Amazon. From coast to coast in the U.S. the landscape is now cluttered with abandoned stores and malls. Amazon has revolutionized retailing.

Let's say a husband wants to by his wife a Chinese jewelry box for Mother's Day. Amazon has dozens of jewelry boxes to choose from—all vividly displayed in color and most available with one- or two-day shipping. Amazon has mastered logistics and is becoming a powerhouse in innovative technology. The Kindle e-reader brought books to the beach, as well as to the bedroom. Alexa will answer questions and turn the TV and lights on and off.

If there is one constant about retail, it's rapid change. The Sears catalogue, or "wish book," contained thousands of items from clothing to food, houses, tools, toys, lumber, appliances, and more. An image of an early Sears catalogue appears below.

Today the Sears catalogue has disappeared and so have most of the firm's retail stores.

Wal Mart has taken over the mom and pop grocery store. The Benton, Arkansas, firm now sells everything from food to hardware. Wal Mart is trying to give Amazon a run for its money in on-line retailing.

Convenience stores are an appendage of gas stations. Their food offerings are seldom nutritious, and food, beverage, and snack items are usually much costlier than prices in supermarkets. Lottery tickets are a big draw and revenue producer.

In retailing, several trends are evident. First, size matters. Amazon and Wal Mart are examples. Next, the heyday of big box stores may be over. They are having trouble competing with on-line retailers. J. C. Penney and Elder-Beerman are examples. Nordstrom and Macy's are feeling the effects of consumers shifting shopping habits. Thirdly, on-line retailing with free shipping and generous product return policies are the trendsetters for today's millennials.

Writer Tyler Durden published a list of endangered retailers in 2017. The list includes: [52]

- Boardriders SA - sporting subsidiary of Quiksilver
- The Bon-Ton Stores - parent of department store chain
- Fairway Group Holdings - food retailer
- Tops Holding II - supermarket operator

- 99 Cents Only Stores - discount retailer
- TOMS Shoes - footwear company
- David's Bridal - wedding dresses and formalwear seller
- Evergreen AcqCo 1 LP - parent of thrift chain Savers
- Charming Charlie - women's jewelry and accessories
- Vince LLC - clothing retailer
- Calceus Acquisition - owner of Cole Haan footwear firm
- Charlotte Russe - women's clothing
- Neiman Marcus Group - luxury department store
- Sears Holdings - owner of Sears and Kmart.
- Indra Holdings - holding company owner of Totes Isotoner
- Velocity Pooling Vehicle - does business as MAG, Motorsport Aftermarket Group
- Chinos Intermediate Holdings - parent of J. Crew Group
- Everest Holdings - manages Eddie Bauer brand
- Nine West Holdings - clothing, shoes and accessories
- Claire's Stores - accessories and jewelry
- True Religion Apparel - men's and women's clothing
- Gymboree - children's apparel

Chances are most Americans have shopped at one or more of these retailers. It reminds one of the rust belt cities where manufacturing departed for venues with cheaper labor – in most cases outside the U.S. The irony here is the fact that firms on the endangered list above obtain merchandise from overseas suppliers.

Grocery shopping is changing. Groceries are being ordered on-line and are loaded into your car—a good way to stop impulse buying. Wal Mart and Kroger are examples of shop and load. Stores will also deliver to your home and place food in your refrigerator.

Restaurants are using Grubhub, DoorDash, Seamless, and UberEats to deliver prepared meals to your home or favorite picnic location.

Delivery.com is another delivery service that goes beyond the menu. You can get lunch, groceries, a bottle of wine, or even get your laundry with this app. Similar to Grubhub and Seamless, Delivery.com doesn't charge you a fee to use its service. Instead, the company makes their money by taking a small percentage of your pre-tip subtotal. [53]

Enterprise will deliver a rental car to your door. Uber is replacing taxi cabs. An Uber driver and car can be summoned with an app on your smartphone and the charge will be automatically posted to your credit card.

Banks are foregoing brick and mortar locations to concentrate on digital banking, Brick and mortar banks are adopting a café atmosphere.

Local and state governments are still promulgating regulations that cast a safety net over growing business enterprises. Auto dealers are being shielded from big auto malls. Some cites have banned Uber to help save their taxi cabs. Driverless cars are forbidden in many locations.

The gateway to the future may be Amazon Go stores. Shoppers use a smartphone app to enter the store. Once they scan their phones at a turnstile, they can grab what they want from a range of salads, sandwiches, drinks and snacks—and then walk out without stopping at a cash register. Sensors and computer-vision technology detect what shoppers take and bills them automatically, eliminating checkout lines. [54]

At the end of the second decade of the 21st century, retailing is increasingly technology driven. Also, today's consumers want things now!

Retail Landscape in 2050

Will physical stores be gone? Will shopping malls cease to exist? Will there be a tsunami in retailing? The answers to these questions are "no." But changes in how we shop and where we shop will be noticeable. Your authors started out with the assumption for once in *The World Beyond Tomorrow*, the changes in 2050 in retailing may not be as dramatic as the thirty-year period between 1990 and 2020 when Amazon and advancements in digital technology revolutionized the retail experience. Forget that assumption. Magic potions lie ahead.

Let's start our journey to 2050 by summarizing what others have envisioned for retailing in the mid-point of the 21st century.

- Augmented reality technology will become the norm, allowing people to walk around stores, meet friends and test products from the comfort of their home. [55]

- People will make all their purchases from home, trying on clothes in virtual reality changing rooms and getting advice from AI (artificial intelligence) shop assistants that know exactly how to cater for their tastes. Online deliveries dropped into the back garden by flying robot drones will become a part of everyday life. [56]

- 3D Printers: Architects around the globe have actually been racing to build the world's first 3D-printed house. In China, a company named Winsun said it built 10 3D-printed houses in one day—each costing just $5,000. A professor at USC is working on a gigantic 3D printer that can build an entire house, with electrical and plumbing conduits. [57] This vignette is included here as selling houses can be considered retailing.

Clearly retailing in 2050 will be driven by augmented reality. This means people will shop from home rather than in a physical store. Once they select their purchases, logistics will come into play to get products from a central location to the purchaser. Drones will replace UPS and Fed Ex trucks. Amazon will have a head start as it already has the infrastructure to adapt easily to A/R purchasing.

Physical stores will still exist for those that want the touch and feel experience. Many of the stores will be cashier-less.

3D printing will have a major impact on the construction and clothing industries. At this point you are probably wondering how a 3D printer works—it's hard to envision. The U.S. Department of Energy outlines the process as follows: [58]

It starts with creating a 3D blueprint using computer-aided design (commonly called CAD) software. Creators are only limited by their imaginations. For example, 3D printers have been used to manufacture everything from robots and prosthetic limbs to custom shoes and musical instruments. Oak Ridge National Lab is even partnering with a company to create the using a large-scale 3D printer, and America Makes—the President's pilot manufacturing innovation institute that focuses on 3D printing.

Once the 3D blueprint is created, the printer needs to be prepared. This includes refilling the raw materials (such as plastics, metal powders or binding solutions) and preparing the build platform (in some instances, you might have to clean it or apply an adhesive to prevent movement and warping from the heat during the printing process).

Once you hit print, the machine takes over, automatically building the desired object. While printing processes vary depending on the type of 3D printing technology, material extrusion (which includes a number of different types of processes such as fused deposition modeling) is the most common process used in desktop 3D printers.

Material extrusion works like a glue gun. The printing material—typically a plastic filament—is heated until it liquefies and extruded through the print nozzle. Using information from the digital file—the design is split into thin two-dimensional cross-sections, so the printer knows exactly where to put material—the nozzle deposits the polymer in thin layers, often 0.1 millimeter thick. The polymer solidifies quickly, bonding to the layer below before the build platform lowers and the print head adds another layer. Depending on the size and complexity of the object, the entire process can take anywhere from minutes to days.

After the printing is finished, every object requires a bit of post-processing. This can range from unsticking the object from the build platform to removing support structures (temporary material printed to support overhangs on the object) to brushing off excess powders.

It's easier to think about 3D printing if you imagine baking a cake using a recipe you found on the web. Put all your ingredients into the printer and your cake will be baked, iced, and decorated to perfection.

Retailing in the World Beyond Tomorrow will be revolutionized by augmented reality and 3D printing. It's hard to visualize the feel, smell, and touch experience in a place, person or object that may be a half world away.

It's even more difficult to grasp the concept of putting building materials into a printer and having the home of your dreams built. Finally, try and get your mind to visualize the retailing environment in 2050. Talk about a Wish Book!

CHAPTER SIX

The Workplace Beyond Tomorrow

To enjoy the necessities of life requires expending a certain amount of energy in what we call work. This can be done outside the home or in the home. The Merriam-Webster Dictionary devotes an entire page to defining *work*. The definitions can best be summed up as: [59]

> **a:** to perform work or fulfill duties regularly for wages or salary: *works* in publishing
>
> **b:** to perform or carry through a task requiring sustained effort or continuous repeated operations: *worked* all day over a hot stove
>
> **c:** to exert oneself physically or mentally especially in a sustained effort for a purpose or under compulsion or necessity

In 2019, two trends are evident. First, much office-based work is moving into the home thanks to computer technology. Secondly, robotics in factories are replacing humans where repetitive tasks are involved. Henry Ford's Rouge plant combined all the resources in one place as workers on assembly lines assembled the Model T Ford. Today just-in-time delivery and robots are the key ingredients in automobile manufacturing.

Today, many companies are undergoing a digital transformation. Artificial intelligence, cloud and mobile, and increased automation have given them the ability to transform every aspect of their business. The use of technology and digitization of every sector of business will have a fundamental impact on 2020 workplace trends. Moreover, management will increasingly turn their attention to the millennial generation, whose attitudes will profoundly reshape society. When compared to older generations, who simply want job stability, younger workers want to have engaging jobs, a purposeful life, an active community, and financial stability. [60]

The Googles and the Apples of the world have changed the workplace. Thousands of highly skilled employees are gathered together on one campus so they can work and brainstorm together. Apple Park looks like a giant spaceship. A screenshot of Apple Park appears below. [61]

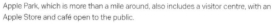

Apple Park, which is more than a mile around, also includes a visitor centre, with an Apple Store and café open to the public.

That's in addition to a 100,000-square-foot fitness centre for Apple employees, 300,000 square feet of secure research and development facilities, and two miles of walking and running paths for employees, underground parking plus an orchard, meadow and pond.

Apple has given us a glimpse of the Workplace Beyond Tomorrow. These are the trends that will carry forward into the workplace of 2050:

- Open spaces with walking and running paths
- Underground parking
- Fitness centers for employees

Robots, augmented reality and quantum computing will reshape how work is done and, most likely, where it is done—on a global basis. Unfortunately, many jobs today will either disappear or be considerably less prevalent in the future. In a YouTube video, Alux.com summarizes the following jobs that will fall prey to technology and artificial intelligence (AI). [62]

1. Drivers
2. Farmers

3. Printers and Publishers

4. Cashiers

5. Travel Agents

6. Manufacturing Workers

7. Dispatchers

8. Waiting Tables and Bartenders

9. Bank Tellers

10. Military – Soldiers

11. Fast Food Workers

12. Telemarketers

13. Accountants and Tax Preparers

14. Stock Traders

15. Construction Workers

We're not going to miss telemarketers, but a disappearing military? Who's going to keep us safe and out of harm's way? Drones appear to be the answer. If you think all of this is depressing, Oxford University researchers predict 47% of today's jobs will disappear in the next twenty-five years. [63] That article was published on Christmas Eve, 2016. Bah! Humbug! [64]

The Workplace in 2050

The key to the workplace at the midpoint of the 21st century is having a job. Robotics and drones will be plentiful and someone, or something, must build them. 3D printers will be capable of assembling widgets and gadgets into a finished product as long as CAD (Computer Aided Design) blueprints are in the recipe book.

A good place to start as we journey into 2050 is too see what other futurists think. We'll start with an article written by Melissa Stanger in 2016. In her interviews, the following comments are relevant: [65]

- Manual jobs are most at risk, as these jobs will be performed by robots. Jobs that require empathy, like social workers and caretakers will be least at risk.
- Employers could start recruiting labor from a global pool of freelancers instead of traditional, full-time employees. This could mean lower pay and fewer benefits.
- People are living longer, and the cost of living keeps going up, requiring many to keep working much later in life. Younger generations also aren't saving money for retirement the way their parents' generation did, because they can't afford it.
- Employees could be monitored. Sensors may check their location, performance, and health.
- More companies will dissolve traditional offices and headquarters and use coworking spaces.
- Driverless cars will make the daily commute faster and easier.

Clearly, the workforce in 2050 will need to develop new skills, so training will be paramount. But lest we think the future is doomed because robots will replace humans, consider the new jobs that we cannot envision today. A short history lesson is in order. Twenty-five years ago, these jobs didn't exist: [66]

- Social media managers and workers
- Big data architects
- App developers
- Bloggers

Also, considers the products that didn't exist, and the workforce that developed them: smartphones, tablets, 3D cameras, drones, etc. Facebook has over

7,000 employees. Apple has over 115,000 employees. The takeaway here is that there will be new jobs in the next thirty years that we can't imagine today.

We discussed training to learn new job skills in 2050. Consider this. Tech of the future: nanobots. By the year 2050, nanobots will plug our brains straight into the cloud, which will give us full immersion virtual reality from within the nervous system. What we do now with our smartphones, we will be able to do with our brains: we'll be able to expand our neocortex in the cloud. And forget about memory problems, evidence problems, etc. [67] A Nanobot is an extremely small robot (a machine controlled by a computer) that can do things automatically. An image of a nanobot appears below.

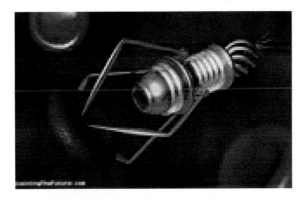

Humans in 2050 will be trained by nanorobots and other forms of artificial intelligence to learn the skills needed to staff the jobs of the future, many of which do not exist today. There will be more leisure time so more people will be employed to service travel and vacation needs. While robots will be capable of servicing much of the hotel industry, the personal touch will be expected by consumers in high end resort areas.

Teaching in the next thirty years will be remote, carrying forward with today's trends to teach on-line. It wouldn't be unreasonable to imagine a complete technological takeover in terms of mode of delivery. Using visual media to help students grasp concepts better is already becoming a huge trend. For sure, the better days of this trend are in front of us. Expecting students to run an AutoCAD simulation in their heads from the (often poorly drawn) blueprint on the chalkboard is the surest way to lose their involvement. To reduce this load on students, teachers will make use of miniature models and simulations in 2050. They will factor in the headway being made in 3D printing technology and students will no longer have to rely on the artistic caliber of their teachers to communicate difficult concepts. [68]

From secondary schools to universities, on-line teaching will mitigate the need for classrooms and sprawling campuses. Augmented reality will teach medical students how to interact with the anatomy of the human body. The use of cadavers as a teaching tool will cease.

Biometrics will be the attendance takers of the future. Want to sleep in late? Your remote teacher will know you're not present in her/his virtual classroom.

Robots will be used in teaching; Autistic students prefer a robotic teacher to a human one. As the age of AI approaches, the question of whether robots can replace teachers looms larger. Anthony Seldon, vice chancellor of the University of Buckingham, claims that robots can never replace teachers because teachers inspire us. But AI can be adapted to each student's individual learning style. [69]

This Photo by Unknown Author is licensed under CC BY-SA-NC

The Workplace Summarized

The workplace in 2050 will feature a kaleidoscope of changes due to technological innovation and artificial intelligence. Robots will continue to replace humans in many occupations. The future is not necessarily bleak. Nanorobots will teach us new skills so we can adapt to the changing landscape. Moreover, there will be new jobs created that do not exist today.

The workplace of the future will be less personalized, but those occupations that require touch and feel, and those that require empathy, will be in demand.

Robots that intereact with people, rather than machines, will take on a human form. They will be what we refer to as cyborgs today: a hypothetical person whose physical abilities are extended beyond normal human limitations by mechanical elements built into the body. An image appears below.

The Workplace Beyond Tomorrow will be an amazing place. If you adapt easily to change, you'll love it.

CHAPTER SEVEN

The Future of Government

If predicting the future of anything is fraught with peril, what the government will be like in the future must be the most perilous of all. The one thing we can say with certainty is that the future will not be predicted by using a straight line from the past. Still, if we do not consider the past, we have no real standing for any guesses at the future. We are regularly counseled to study the past in order to avoid the mistakes made then. So, we begin with some thought to the past.

We also must consider what features of government are worthy of our attention. It is probably not worthwhile to consider exact tax rates, for example, since they change at about the same rate as light bulbs burn out. But we can usefully consider if the general tendency is to increase or lower the cost of government. We should ask whether the government has imperialist tendencies, or whether citizens see government leadership as serving the public, rather than ruling it. We can consider whether a government will be more or less likely to try to resolve international conflicts with force or negotiation. An important question to ask is whether the government will expand or restrict the right to vote. Still another important point is how the government will treat the weakest persons who live in the country, people like non-citizens, children, minorities, or women.

We will consider only the government of the United States. In this country, we enjoy the benefits bestowed on us by the Bill of Rights, the first ten amendments to our Constitution. Those include:

1. Freedoms of religion, speech, press, assembly; freedom to petition the government and to keep and bear arms; and freedom from quartering soldiers.

2. Protections provided by search warrant, trial by jury, double jeopardy, the right not to testify against oneself, due process, taking of private property, speedy trial, confrontation of witnesses, assistance of counsel, bail, and cruel and unusual punishment.

3. Those not listed are the states' or the people's rights.

How will the government of the future honor these?

Let us begin with voting. The Constitution certainly introduces the idea of voting but allows the states to decide who shall vote. Voting was almost exclusively restricted to white men of property. Those voters chose members of the House of Representatives and state legislatures, but Senators were chosen by each state's legislature. The individual voters elected representatives to pick the President. Those representatives are the Electoral College.

In 1913, the Constitution was amended so that Senators were elected by popular vote, but the Electoral College is still with us. The use of the Electoral College has led us to elect at least four Presidents for whom most people did not vote. This may seem odd, but those presidents have been accepted as President in the same way as those who won both the Electoral College and the popular votes. So to our first prediction: by 2050, the Electoral College will have been neutralized, most likely by states deciding to use a system of proportional choice of electors, rather than the winner-take-all system that is used now.

State senators were elected to represent fixed land areas, such as a county at one time. The U. S. Supreme Court ruled that State senate districts should be about equal in population, thus ending the representation of land that had been the case earlier. (Reynolds v. Sims, 377 U.S. 533 (1964), 1964) This is another example of the general trend to honor the idea that every person's vote should count the same.

The U. S. senators, of course, are chosen according to the state they represent regardless of the population of that state. California has two senators, just as Wyoming does, although California (19 million per senator) has many more people than Wyoming (quarter million per senator) does. The number of members of the House of Representatives that each state has is decided according to the population of that state, with the exception that each state has at least one member of the House. In the case of California (seven hundred thousand per person) and Wyoming (568,000 per person), again, the difference is real, and this is not even close to the largest differences (Berg-Andersson 2017). While this is a vote that disproportionately values individual voters, it is enshrined in the U. S. constitution and arose from the idea at the founding of the country that individual states were sovereign and therefore were worthy of representation as a state. Therefore, we predict this will not change, at least not by 2050.

The U. S. government spending, when adjusted for inflation, was relatively constant, except for in times of war, from the founding of the country until about 1930. The increase from 1930 to about 1980 is the result of the depression and World War II. From 1980 on, government spending looks like a binge. Historically, we spent about 3% of Gross Domestic Product (GDP) except during crisis times like war. Since 1980, the comparable number is 20% or about six times as much in constant comparison. Much of that increase in spending was financed by loans, increasing the national debt greatly. Because of the debt, there is every reason to suppose that spending will increase whether wars come or go, or whether government services increase or decrease. This is because the ballooning national debt will require interest payments at levels that will require tax increases, spending cuts, or refinancing to continue to pay just the interest on the national debt. So, our next prediction is that government spending will increase, and the national debt will increase even more than spending does (Metrocosm 2016).

At the birth of the nation, the U. S. had little interest in expansion, but that soon changed. Perhaps the biggest expansion of territory for the country was the Louisiana Purchase in the early 1800s. From that time to about the mid-1900s, our country was truly an imperialist expansionist nation. Certainly, the acquisition of property as a result of the Mexican or Spanish American Wars fit that model. Some would say that the annexation of Texas was engineered to carry forward such a policy. Manifest Destiny was a philosophy that espoused the idea that the U. S. had a destiny to expand to the West Coast and that did happen. In the twentieth century, two territories of the U. S. (Alaska and Hawaii) became part of the country as states. On the other hand, the Iraq war could be seen as an effort to secure American oil interests. It seems that we are not so solidly imperialist-expansionist as we have been in the past. Neither are we so opposed to colonialism as we might like to think we are.

On the other hand, populations around the world are moving towards self-government more than they have in the past. This will make our efforts at expansion harder. These competing forces also make it harder to predict the path of government policy in this area. Still, this book is about the future and so we will predict that the same quagmire of uncertainty in our foreign policy will continue.

Our next question is whether our government will serve our people or rule them. We think the founders intended for our leaders to be servants of the people. Over time that role has eroded so much that, even in the USA, our leaders prefer to rule us. This trend seems irreversible. By 2050, we will be accepting government roles that today would seem totalitarian. For example, our government will probably use some draconian measures to restrict the burning of hydrocarbons for energy, as a way to stem climate change. The rule changes will be gradual, and we will become used to the new paradigm without even noticing it has changed. There will be, of course, the few on the fringe of our society calling out the change and the way government operates, but most of us will ignore them.

Will our future hold more negotiation or war? This is probably best left to the war chapter. Here we will only say that the governments of the world will seem to do more negotiation, but much of it will be superficial. Its real purpose will be to fend off the opponent until the negotiator's situation improves.

Those items mentioned in the Bill of Rights will be handled differently. Freedom of, or more exactly, freedom from religion will be less respected. First, Fundamental Christians in the U.S. have gained more voting power over the last several years. They have also become more vocal about their feelings of oppression. As a result, Christians will be able to exercise their religion at public functions. Two cases on point are the public official in Kentucky who did not want to issue licenses for marriage to same sex couples and the bakery that did not want to work on a cake for a marriage between two members of the same sex. On the other side, the continuing struggle with Muslim based religions or sects and the countries where they are a majority will create ill will towards them in the United States and some aspects of their religion will be restricted. Specifically, facial concealment in public will be strictly curtailed. Women's rights will be protected under US law, more so than they are now.

The right to keep and bear arms will be curtailed by some licensing system. It will happen because the defenders of the Second Amendment on weapons will hold out for no changes for a long time. Eventually a truly horrific tragedy, far eclipsing anything we have seen so far will occur. It will be seen as preventable with changes in the right to keep and bear arms legislation. But the public will be so angry when this occurs, that changes that would have been very mild and very acceptable to persons opposed to the right to keep and bear arms in 2019 will not be acceptable to the anti-gun groups in 2050. The changes they will require in 2050 will be draconian. Gun owners will be required to register all guns. No handguns will be allowed. Long guns will be kept at a police facility and will only allowed to be retrieved with advance notice. Ammunition will be available only through law enforcement. Background checks and drug testing will be required of all persons who possess a gun, even registered collectors.

The rest of the freedoms enumerated in the Bill of Rights will all be diminished, especially as we see Muslims as a danger to our society. Increasingly we see the Press as an enemy of the people and the government will try to limit the freedom of the press. These restrictions will be supported much as they were during World War II, when we saw the nation in peril, and we were willing to accept curtailed freedoms for the war's duration. The struggle over Islam will go on so long that the restrictions will begin to seem normal and so will continue into the future.

These predictions may seem dire. We think we are generally optimistic and that most Americans will be comfortable with the changes. The US will still be the greatest nation on earth, as it is now.

CHAPTER EIGHT

Finances Beyond Tomorrow

How do we, or the government, manage money and what about banking and investment? Will paper money and coinage disappear as we get closer to the mid-point of the 21st century, and virtual money such as bitcoin be the norm? Will the dollar, pound, mark, yen, peso, dinar, rupee, yuan, franc, krone, lira, shilling, and euro become extinct?

Bitcoin hit the currency market less than a decade ago. Bitcoin is a digital asset and a payment system invented by Satoshi Nakamoto, who published the invention in 2008 and released it as open-source software in 2009. The system is peer-to-peer; users can transact directly without an intermediary. Transactions are verified by network nodes and recorded in a public distributed ledger called the block chain. [70] As of June 2019, there were nearly 18 million bitcoins in use. The generic name for this new electronic currency is cryptocurrency. An image of a Bitcoin appears below.

There are several problems associated with cryptocurrency. These include: [71]

- It's still extremely tech-y. It's hard to grasp the concept of something you can't see or touch.

- The scams are plentiful. Are you investing to purchase Bitcoins or sending your money to a scammer or thief?

- The price volatility scares away most people who aren't interested in a get-rich-quick concept. The price went from $20,000 to $2,000. Today it's $6,000.

- There's also the difficulty that comes with spending your cryptocurrency. Many merchants won't accept cryptocurrency because of price fluctuation and the high fees to accept it. Credit cards are now available that convert electronic money into regular money.

- There's also a major threat of government regulation choking it out. Several nations have already banned the currency.

With electronic money, you don't know who holding your money and it's not insured by anyone. However, in the next thirty years, global trade will demand a global currency that has a stable market value.

Paper money will disappear over time. In Sweden, only 2% of transactions are in paper or coins. The image below is courtesy of Getty Images.

Look for the European Union to transcend to an entirely cashless society. Today there are 28 countries in the E.U. Seventeen of these countries use the Euro as their official currency. [72] These seventeen will take the paperless journey to electronic currency.

What will the U.S. and China do? China has begun hording gold. U.S. currency is still the most highly regarded currency in the world. Many economists consider the $100 bill as a proxy for foreign demand. A 2018 research paper estimated that 80 percent of $100 bills were in other countries. Possible reasons included economic instability that affect other currencies and use of the bills for criminal activities. [73] It is unlikely the U.S. will adopt an electronic currency until forced to do so by global pressure.

Is China hording gold to tie the value of the Yuan to the price of gold? The U.S. abandoned the gold standard in 1971. The chart below reflects the holding of gold by various countries. Although the amount of gold held be China appears small compared to the U.S., China has recently stepped up its gold purchases.

China is simply its holdings. Viewed as a percentage of all foreign exchange reserves, China's gold reserves remain tiny—about 1.6 percent, as shown above. Compare that with 73 percent for the U.S., 67 percent for Germany and 65 percent for France and Italy. Given the size of China's economy, it *should* own more gold, just like the other big players in the global economy. By increasing its gold reserves, and reducing its holdings of U.S. dollars, China is spreading its risk and reducing its volatility. [74]

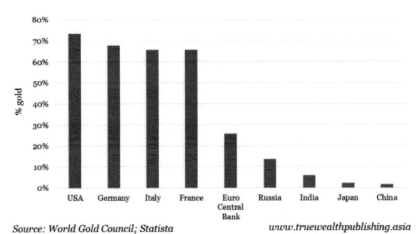

Gold as Percentage of Foreign Exchange Reserves

Source: World Gold Council; Statista   *www.truewealthpublishing.asia*

China has banned the use of cryptocurrency. As we move toward 2050, the U.S. and China will keep the dollar and Yuan, respectively. That said, both countries will open the door to the use of electronic currency, but it will be highly regulated.

Other Financial Considerations

Finances cover more than just currency. The movement of goods and services around the globe impacts national economies. China is an example of a tightly controlled nation that has embraced state-run capitalism.

Fintech is the current buzz word in the financial world. Fintech now describes a variety of financial activities, such as money transfers, depositing a check with your smartphone, bypassing a bank branch to apply for credit, raising money for a business startup, or managing your investments, generally without the assistance of a person. [75] As the mid-point of the 21st century approaches, brick and mortar banks will become a rarity. Credit cards, mortgage lenders will become obsolete. Consider these on-line providers cited in the latest footnote:[76]

- Bitcoin – The cryptocurrency wave.
- Affirm – Seeks to cut credit card companies out of the on-line shopping process.
- Better Mortgage – Streamline the home mortgage process and obviate the need for a home mortgage broker.
- GreenSky – home improvement loans with a zero-interest promotional period.
- Tala – Microloans for those with poor credit in developing countries.
- Upstart – Loan originator that hopes to replace FICO scores by using different data sets to determine creditworthiness.
- LendingHome – Bridge loans for house flippers.
- Ellevest – Recognize women live longer and have different salaries and less time for savings to grow.
- Robinhood – Charges no fees for mobile stock trades.
- Kabbage, Lendio, Accion and Funding Circle (among others) offer startup and established businesses easy, fast platforms to secure working capital.

Big banks are scrambling to catch up as the on-line financial invasion picks up steam. But you cover one base and a new start up emerges. When these new entrants get into the merger and acquisition field, the bells will start to toll for the world's mega-banks.

Global Growth

Cheap labor and plentiful natural resources are the key ingredients for growth, both in 2020 and in 2050. Growth is also maximized as manufacturing replaces agriculture. Price Waterhouse Cooper (PWC) has graphed growth in the world's top economies. The results are shown below. [77]

Emerging markets will dominate the world's top 10 economies in 2050 (GDP at PPPs)

	2016	2050	
China	1	1	China
US	2	2	India
India	3	3	US
Japan	4	4	Indonesia
Germany	5	5	Brazil
Russia	6	6	Russia
Brazil	7	7	Mexico
Indonesia	8	8	Japan
UK	9	9	Germany
France	10	10	UK

☐ E7 economies ☐ G7 economies

Sources: IMF for 2016 estimates, PwC analysis for projections to 2050

Vietnam, the Philippines and Nigeria could make the greatest moves up the rankings

	Vietnam	Philippines	Nigeria
2050	20th (up 12 places)	19th (up 9 places)	14th (up 8 places)
2016	32nd	28th	22nd

Average annual GDP growth rate, 2016-2050

| 5.1% | 4.3% | 4.2% |

Sources: IMF for 2016 estimates, PwC analysis for projections to 2050

More troubling is the fact that the U.S. is falling behind China and India as a percentage of the world's GDP. This is reflected in the following chart. [78]

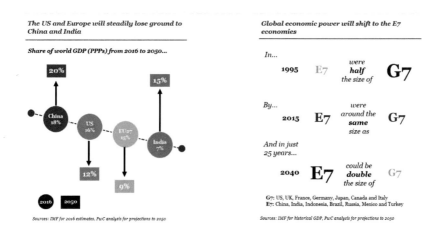

The US and Europe will steadily lose ground to China and India

Share of world GDP (PPPs) from 2016 to 2050...

Sources: IMF for 2016 estimates, PwC analysis for projections to 2050

Global economic power will shift to the E7 economies

In...

1995 — E7 were half the size of G7

By...

2015 — E7 were around the same size as G7

And in just 25 years...

2040 — E7 could be double the size of G7

G7: US, UK, France, Germany, Japan, Canada and Italy
E7: China, India, Indonesia, Brazil, Russia, Mexico and Turkey

Sources: IMF for historical GDP, PwC analysis for projections to 2050

The U.S. has fallen from capturing 16% of the world's GDP to 12%, while China has increased from 18% to 20%. The data comparing the G7 with the E7 shows the increasing strength of the Asian countries in capturing GDP market share.

CHAPTER NINE

Climate Beyond Tomorrow.

F our feet of hail in Guadalajara, Mexico's second largest city. Rains and flooding in the mid-western section of the U.S., destroying a season of crop planting. Tsunamis, volcanos, forest fires, and earthquakes. Melting icebergs in the Antarctic. Climate change is more than a political question. Mother nature does strange things, but so do humans. The environment is fragile, and mankind has tampered adversely with its eco-system—and continues to do so. An image of the hail in Guadalajara appears below.

National Geographic recently published an article on climate change. Temperatures are rising at twice the rate they were 50 years ago. The culprits are greenhouse gases, including carbon dioxide, methane, and nitrous oxide. Most come from the combustion of fossil fuels in cars, buildings, factories, and power plants. The gas responsible for the most warming is carbon dioxide, or CO_2. Other contributors include released from landfills, natural gas and petroleum industries and agriculture (especially from the digestive systems of grazing animals); nitrous oxide from fertilizers; gases used for refrigeration and industrial processes; and the loss of rainforests that would otherwise store CO_2. [79]

In June 2019 CO2 levels reached 412 ppm (parts per million) in the atmosphere. The chart below traces CO2 levels back 400 million years ago. The CO2 levels in the last 2 glacial cycles are reconstructed from ice cores. [80]

Agra

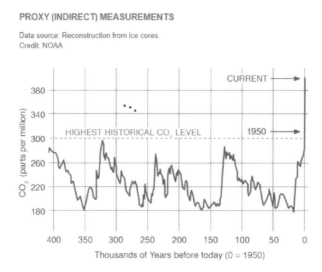

A group of meteorologists has shown that a decrease in solar radiation and an increase in volcanic activity will cause temperatures to plummet. This is reflected in the graph below. [81]

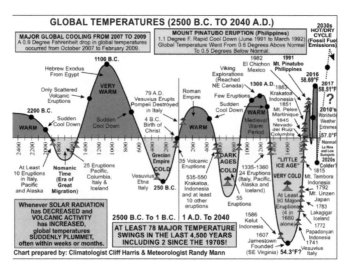

In the latter half of the 18th century, the beginning of the industrial revolution brought faster and easier ways to manufacture goods and revolutionized farming. Into the 19th century, belching smokestacks dotted big city landscapes. As cars rolled off assembly lines and took to the roads, their emissions polluted the atmosphere. Armament manufacturing from two world wars compounded pollution problems. An image of Ford's River Rouge plant appears below.[82]

In 2020, there is a world-wide consciousness that our fragile eco-system must be preserved. Governments, individuals, and groups have spoken with a collective voice. There are those who disagree, but in time the evidence will become so overwhelming that minds will change, and a "just do the right thing" mentality will prevail. Moreover, most of the damage has already occurred.

European countries lead the way in protecting the environment. The graph below reflects ongoing efforts to protect mother earth. [83]

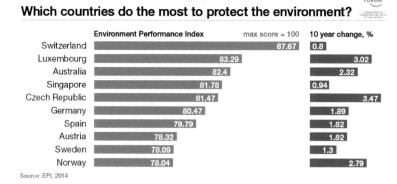

Which countries do the most to protect the environment?

	Environment Performance Index max score = 100	10 year change, %
Switzerland	87.67	0.8
Luxembourg	83.29	3.02
Australia	82.4	2.32
Singapore	81.78	0.94
Czech Republic	81.47	3.47
Germany	80.47	1.89
Spain	79.79	1.82
Austria	78.32	1.82
Sweden	78.09	1.3
Norway	78.04	2.79

Source: EPI, 2014

In Switzerland, a carbon tax is placed on greenhouse emissions. Special legislation applies to housing density in large cities and water pollution stations help keep lakes and rivers clean.

The University of Notre Dame developed an index on the countries best prepared for climate change. As can be noted from the graph below, the U.S. is ranked as the 10th best prepared. [84]

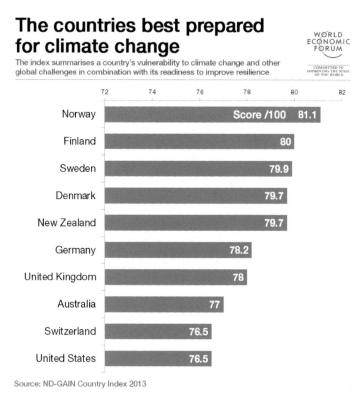

The countries best prepared for climate change

The index summarises a country's vulnerability to climate change and other global challenges in combination with its readiness to improve resilience

Country	Score /100
Norway	81.1
Finland	80
Sweden	79.9
Denmark	79.7
New Zealand	79.7
Germany	78.2
United Kingdom	78
Australia	77
Switzerland	76.5
United States	76.5

Source: ND-GAIN Country Index 2013

Asian countries are noticeably absent from the best prepared list, yet China and India have the fastest growing economies. There appears to be an inverse correlation between economic growth and environmental protection—yet Germany is an example where both growth and protection can co-exist.

Climate in 2050

Will humanity pull itself up by the bootstraps and save mother earth from a climatic disaster by 2050, or do we have to look elsewhere in the galaxy for suitable habitats?

In previous chapters, we discussed changes that will be more environmentally friendly. Driverless cars have fewer parts and will be fueled by solar cells. The same for buses. Trains will travel underground and be fueled by solar cells. Biochemistry advancements will be responsible for developing organic fertilizers. 3D printers will manufacture clothing, furniture, and housing. Moreover, these changes will come well before 2050.

There are forecasts that humanity will cease to exist by 2050 due to global warming. A new report by Australian climate experts warns that "now represents a near- to mid-term existential threat" to human civilization. In this grim forecast—which was endorsed by the former chief of the Australian Defense Force—human civilization could end by 2050 due to the destabilizing societal and environmental factors caused by a rapidly warming planet. [85]

Dire predictions make good news coverage, but one has to look under the hood to see if any parts are missing. What's missing in the above prediction is the effect of technology on the future of climate change.

Geoengineering is an emerging technology. Geoengineering is sometimes called "planet hacking" because it uses a literal hacking of the planet's resources to find new solutions. It's based on the belief that climate change can be halted using man-made means. It usually takes the form of two things: carbon dioxide reduction, like building algae farms, planting trees, capturing emissions from power stations for fuel; and solar radiation management, like releasing volcanic ash as a coolant, arranging mirrors in space to redirect solar rays, or painting roofs white instead of black. It's controversial because we don't know the environmental or health effects of most of these ideas. [86]

In a *World Economic Forum* article, author Alex Gray describes five ways technology can be used to mitigate climate change. These include: [87]

- Power Generation – A Canadian company aims to be the first in the world to create a commercially viable nuclear-fusion-energy power plant. Fusion produces zero greenhouse gas emissions, emitting only helium as exhaust.

- Transport - Researchers at the University of Surrey say they have made a in this regard. They say they have discovered new materials offering an alternative to battery power and proven to be between 1,000-10,000 times more powerful than the existing battery alternative, a supercapacitor.

- Food - One of the alternatives is to start producing lab-grown meat, and to produce meat substitutes that look, taste and feel like the real thing. It might seem like the stuff of science fiction, but companies and investors alike are taking it very seriously. The company, already supported by Bill Gates, has created the world's first meat burger that is entirely plant based. It's made mostly from vegetable protein found in peas.

- Manufacturing - Making the things we use every day puts an enormous strain on the climate. But what if we could take those CO_2 emissions out of the air? There is a Canadian start-up which is working on exactly that: taking carbon dioxide directly from the atmosphere and then using it to produce fuel.

- Housing - Sidewalk Labs is studying homes to make homes in cities more efficient.

The world will be better in 2050, as the air we breathe will be cleaner and the environment more eco-secure, Perhaps the following quote sums it up best. *"We see global warming not as an inevitability, but as an invitation to build, innovate and effect change; a pathway that awakens creativity, compassion, and genius. This is not a liberal agenda; this is not a conservative agenda—this is a human agenda."– Paul Hawken, Author of*

CHAPTER TEN

War Beyond Tomorrow

As we consider the future of war, it would seem natural to consider the history of war, even though that history is not completely outlined, at least at the very beginning of our human existence.

It is safe to say that war has always been a part of our history. Whether it is learned behavior or inherited is harder to say. It is true that practically all known societies practice or have practiced some form of warfare. The experience of war for some societies other than our own are displayed here as the percentage of male deaths caused by warfare. This shows very definitely that the warfare of societies other than our own is much bloodier when including the number of participants. In other words, a Jivaro male is more likely to be killed in warfare than is a male from Europe or the U. S. [88] It would be hard to convince a member of the armed forces of World War II that this is true, since so many were killed or maimed in that war, but it is so.[89]

Percentage of male deaths caused by warfare

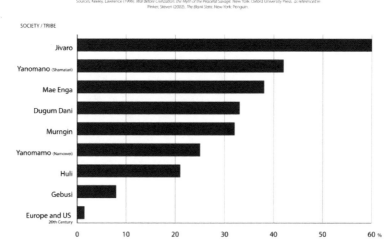

Sources: Keeley, Lawrence (1996). *War Before Civilization: the Myth of the Peaceful Savage.* New York: Oxford University Press, as referenced in Pinker, Steven (2002). *The Blank Slate.* New York: Penguin.

Not only is our society less bloody than the primitive, gentle society we have been led to believe was peaceful, but we have also been trending to less death from warfare since World War II. In fact, at no time in history has such a long period of time passed without war between major powers.

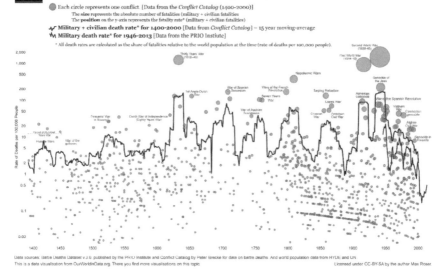

If we are to predict the future of war, it would be good to try to understand the reasons for war. A famous Nazi said:

"Naturally, the common people don't want war; neither in Russia nor in England nor in America, nor for that matter in Germany. That is understood. But, after all, it is the leaders of the country who determine the policy and it is always a simple matter to drag the people along, whether it is a democracy or a fascist dictatorship or a Parliament or a Communist dictatorship.... The people can always be brought to the bidding of the leaders. That is easy. All you have to do is tell them they are being attacked and denounce the pacifists for lack of patriotism and exposing the country to danger. It works the same way in any country." —*at the Nuremberg trials, April 18, 1946*[93]

This quote illustrates one of the reasons we will use in our discussion. People must believe that their interests are being threatened. They also must think that the situation does not allow compromise. Finally, each side must believe that it has a reasonable chance of winning the war and protecting the threatened interests.

Until the American Civil War, it was possible to view war as a glorious struggle between right and wrong with little concern for the reality of war. But photography, with its graphic images of the brutality of every war, began to change that. Add the immediacy of telegraphic results listing the casualties and the willingness to enter into such a bloody escapade with any reasonable hope that winning any war would be worth the human costs began to fade. The threatened interests no longer seemed so interesting.

World War I added to the case against war, with motion pictures of the devastation, although in black and white. Again, reports of casualties were immediate and personal, making any war outcome, win or lose, less attractive. What did it matter if your country won the war but you lost your son? It was even worse to see that he died from an anonymous bullet while lying in a mudhole. The last scene in the film *All Quiet on the Western Front* illustrates that death vividly and add the uselessness of the young man's death.

Now comes World War II, with even more vivid pictures and movies and even more rapid and personal casualty reports. This war ends with the destruction of much of Europe and Russia and the annihilation of Hiroshima and Nagasaki with the atomic bomb.

At this point, it became very hard to argue that a war could be fought to protect a country's interests without destroying those interests, whether the war was won or lost. We believe that is a very large part of the reason that since 1945 there has been no conflict between major powers. This is the longest time in recorded history that has occurred.

The globalization of the world has also played a role in preventing major conflicts. We buy goods from other countries in order to save money on the purchase of children's toys or family clothes or even groceries! We pay for those things when we sell them cars or corn or college education. Therefore, we see fewer things that cannot be compromised because we know the people on the other side so very well.

Also, because we work so closely with other countries on the trade side of our lives it becomes vastly easier to discuss issues and compromise to avoid violent conflict. A trade war is very unpleasant, but it does not involve shooting. And that seems to be a good thing.

So, we predict that wars of such violence as the three just mentioned will not happen again before 2050 and probably long after. Our prediction refers to full-scale war between major powers. Major powers will continue to war with each other, but the wars will be conducted as cyberattacks on voting, defense, infrastructure, and monetary systems. For example, the US may try to break the Internet for Russia, or Russia may try to disturb US voting processes. Both sides will deny the accusation and will continue to do it but in more anonymous ways.

Wars such as those we now have, in which groups of people in a given country will try to overthrow the government, will continue. Those groups who are seen as advancing a major power's vital interests will be supported by that major power. Groups who are opposed to the major power's vital interests will be supported by some other major power. An example might be if the US thought that rebels in Botswana would help with our diamond trade if the rebels ran the government. Russia might think that they would like that diamond trade and so might induce South Africa to support the government of Botswana in opposition to the rebels. We should expect this asymmetric warfare to go on more or less constantly over the next thirty years.

The weapons and outcomes of cyberwarfare are already among us. They include infrastructure disturbance and election interference. The attackers and defenders will use ever more sophisticated computers stationed at ever more remote locations. One weapon system of the future that is being tested is the use of the electromagnetic pulse (EMP) to destroy, at least temporarily, electronic control systems. These weapons seem to offer the advantage of disabling an opponent's electronically controlled weapons without killing or wounding the humans using the system. In the George Clooney movie *Ocean's Eleven*, such a device is used to shut down the security systems in Las Vegas for a short time. This allows the crooks to rig the system so they can complete their theft. There are some flaws in using such a system as a weapon of war. First, electronic systems can be protected from the pulse. Second, the EMP must get pretty close in order to be effective. Finally, the difference between the power of an EMP that will shut down electronic systems and one that will harm humans is very small. That probably means that EMP as a weapon is for the future. But that is what we are writing about.[90]

While the weapons and the warriors will be more in number, they will be doing about the same things as they are now. This will certainly be more inconvenient than it is today, but not more deadly.

Asymmetric warfare, on the other hand, will become much bloodier and nastier. Weapons will become more tightly aimed and more lethal when fired. Military intelligence will spot targets more precisely and correctly and defenders will try to hide behind large populations of noncombatants. This kind of warfare will be endemic in any part of the world where economic inequality is important.

Anyone who can solve any part of the asymmetric warfare dilemma will deserve the highest awards available in our society. In addition, that solution will make our futures much rosier.

CHAPTER ELEVEN

Faith beyond tomorrow

According to Gallup, seventy one percent of Americans claim to believe that that the Bible is either the literal or inspired word of God. According to a Barna Group poll, an estimated 41 percent of Americans over 18 agree that the world is currently living in the "end times." The end times is a reference to a period predicted by the Bible that leads up to and includes the return of Jesus to establish a global kingdom on the earth.

This data shows that a biblical view of the future is a relevant topic to most Americans. Therefore, in this chapter we explore seven predictions the Bible makes about the future. The Bible itself has a disclaimer that no one knows the exact timing of these events. They could occur in the next few years or beyond. However, the Bible does say that there will be indications that the time is getting near. We will explore some of those predicted indicators and you can decide.

If you believe this is all foolishness, increasingly you are not alone. In the past 40 years, the number of people who no longer believe the Bible is relevant has almost doubled from 13% to 25%. In addition, this trend appears to be accelerating.

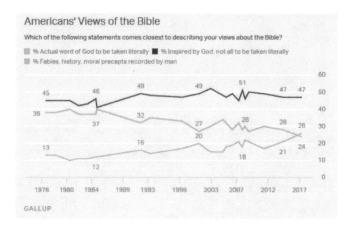

Of course, you can choose to skip this chapter. Choice is what makes life so interesting. We get to choose what we believe but we *do not* get to choose the consequences of what we believe. Go to University A or University B. Marry person A or B. Wear seat belts or not. Spend or save. The choice is yours.

The Bible puts forth a premise. The premise is that God created everything, including us, and that there is a plan for this universe that is unfolding. It also states that He designed each of us with the ability to choose whether to have a relationship with Him or not. It also states that the conclusion of this plan is the establishment of an eternal kingdom where all those in history who have a trusted relationship with Him will be with Him in this kingdom, and those that don't won't. The final premise is that God caused the Bible to be written as a revelation of His purposes and plans. Moreover, in this revelation, predictions about the future are made. In this chapter, we will highlight seven.

Prediction 1: Increase in Travel

About 2500 years ago, a prophet, Daniel, received a strange message from God that predicted that in the "time of the end many will rush here and there." At the time, travel was on foot or using camels or donkeys with a potential range of 20 to 30 miles per day. What an unusual and specific prediction that large numbers of people would be moving about quickly! After 2500, years the significance of this prediction becomes clearer.

In terms of world history, just 110 short years ago, the automobile began to be mass-produced. It is reported that as of 2009 there were 1.4 *billion* cars and commercial vehicles in the world. https://rfidtires.com/how-many-cars-world.html.

In the US alone in 2016, drivers in cars, trucks, minivans and SUVs put a record 3.2 trillion miles on the nation's roads, up 2.8 percent from 3.1 trillion miles in 2015. That is the rough equivalent of every man, woman, and child in America driving from New York to San Francisco *3 times*.

https://www.npr.org/sections/thetwo-way/2017/02/21/516512439/record-number-of-miles-driven-in-u-s-last-year.

Cars on Highway 101 in Los Angeles on Tuesday. Around the nation, Americans drove a record number of miles last year.
Justin Sullivan/Getty Images

Now consider air travel. In 1986, 842,000 passengers were carried by the world's airlines. In 2017, this had grown to 4 billion—almost a five-fold increase in 30 years. In addition, looking at the graph the trend is accelerating.

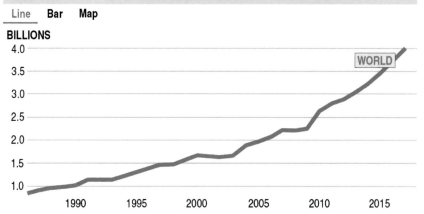

In retrospect, this prophecy seems remarkable. For thousands of years, travel did not change that much and yet in the past 120 years we have gone from traveling at the speed of horseback and a few miles a day to travelling at 500 miles in an hour. Although the trends continue to accelerate, it sure appears this is one prediction that has already been fulfilled. And it is a pre-condition to being in the 'time of the end'.

Prediction 2: Increase in Knowledge

The message that Daniel received 2500 years ago included another prediction. The second prediction was that in the "time of the end knowledge would increase." At the time that must have seemed like a mysterious prediction. Knowledge had been and is always increasing. What did the prediction mean? Well just as time have revealed a better understanding of what the travel prediction meant, so time is unveiling what the increase in knowledge means.

A stunning and unprecedented development in human history is the appearance of the Internet. A convergence of computing technology, wireless technology, and mass production of devices has exploded in terms of world impact. In addition, it has occurred in the blink of an eye when considering human history. In less than 30 years, it has gone from prototype to worldwide availability. Never has such a large-scale device been deployed so quickly.

According to a Forbes article, by 2020 there will be more than 6.1 billion smartphone users. (The population is currently only 7.6 billion.) In addition, by 2020 it is forecast that there will be 50 billion devices connected to the internet.

The amount of data being created is beyond our ability to grasp. More data has been recorded in the past two years than in the entire history of humans. In 2020, 1.7 million characters of data will be created for every human being on the planet *every second.* Everything is being digitized and recorded—phone calls, CT scans, pictures, blueprints, emails, every online purchase, and banking transaction. The list is endless and is the equivalent of a 750-page book being written for every person every second.

The cumulative digital storage by 2020 will be 44 trillion gigabytes. There are over a billion Facebook users every day sending 31 million messages and viewing 2.8 million videos every *minute*. 300 hours of video are uploaded into YouTube every minute. Over one trillion photos are being taken every year, 80% of them on cell phones.

One indication of the significance of this trend is the impact on energy consumption. At current rates of increase, consumption by 2040 computers would consume all of the world's energy production. Of course, this won't happen because new energy-saving technologies will be developed and new sources of energy will be developed. However, energy consumption is one indicator of the geometric growth in creating and accessing information.

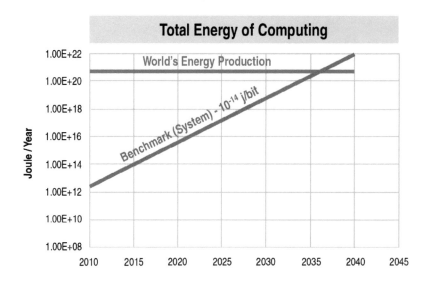

This explosion of information and the rate at which it has occurred is beyond our ability to comprehend. However, even more remarkable is that this broad trend was predicted 2500 years ago at a time when we had no idea what it meant. This is another indicator that is a precondition of the "time of the end."

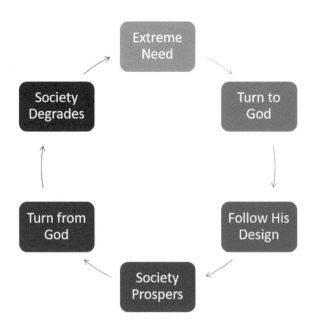

Prediction 3: Decrease in Faith

One sort of irony predicted in scripture is that nations that live according to the principles God prescribes for society prosper and the prosperity itself leads to less faith. The cycle described is one where extreme need causes people to seek God and be willing to follow His design for life. As societies live according biblical principles, society works and people prosper. Prosperity, however, reduces the sense of need for God and people abandon their relationship with Him and His design. Society then ceases to work well; it degrades and eventually it returns to extreme need.

Where is the US in this cycle? Tobin Grant (a political scientist) has compiled a "religiosity" index based on 400 survey results over the past 60 years tracking church attendance, attitudes towards prayer and the Bible, etc. Because the index is a composite of many factors, it does not show absolute values on the left axis but a sense of relative change over time compared to the average.

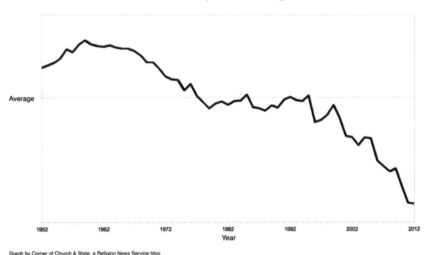

The Great Decline: 60 years of religion in America

Graph by Corner of Church & State, a Religion News Service blog
Source: Aggregate Religiosity Index. J. Tobin Grant. *Sociological Forum.*

It shows that after World War II there was a peak, then a steady decline in the 60s and then relative stability in the 70s and 80s, and a precipitous decline beginning in the early 90s. (The proximity to the bottom of the graph does not mean we are at zero faith. It just means we are at two standard deviations below the average.)

The Bible clearly predicts what happens to societies that move away from His design for community. People begin to lose their ability to think clearly about what works and what does not in life. It predicts an abandonment of the core principle on which society is built—the family. In God's design, healthy societies are built on reproducing the next generation through a stable family unit of one man and one woman committed to each other for life. This commitment itself is rooted in the belief in a sovereign God who knows what is best for a fulfilled life and healthy society.

The Bible predicts that as people cease to believe Him, they abandon the sanctity of marriage and the purpose of human intimacy itself. Let's review what the data says about this prediction.

In the US, the number of babies born outside of wedlock has increased from about 80,000 per year in 1950 to 1.6 million in 2010. In terms of percent of births, it has gone from about 3% to 41%.

Out-of-wedlock childbearing has risen over the past several decades

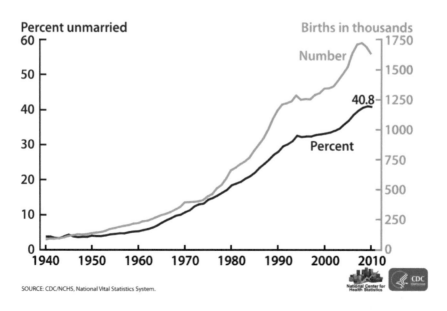

SOURCE: CDC/NCHS, National Vital Statistics System.

A related and a far more dangerous trend is the loss of the *belief* that this is a problem. In just 13 short years, the belief that this is a problem has dropped from 50 to 35%. What this means is that people are abandoning God's view of how society is designed to work. As indicated in the introduction, He has given us the choice to decide that. But we do not get to choose the consequences.

Americans' Views of the Morality of Having a Baby Outside of Marriage

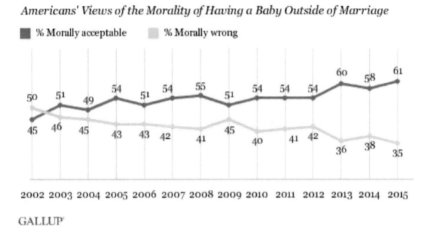

GALLUP'

Another predicted outcome of abandoning God is the view of sexuality itself. The Bible predicts that as societies move away from God, they begin to reason that homosexuality is ok. The description in the Bible is that they cease to be able to see the obvious, to reason correctly. Again, let's see what the data says about this in America.

In the past 15 years, people who have no affiliation with faith have moved from 60% approval of same sex marriage to 75%. More surprising however, is that Evangelical Protestants have moved from 12% to 27%. The rate of approval among those who claim to have faith in God has more than doubled.

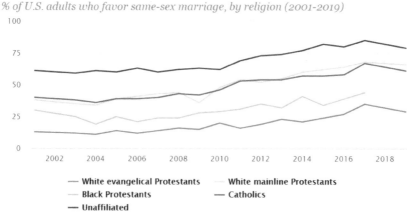

% of U.S. adults who favor same-sex marriage, by religion (2001-2019)

From a "reasoning" aspect, scripture is predicting that we as a society will cease to recognize the obvious, that male and female reproductive organs were designed to work together for intimacy, pleasure, and raising the next generation of people faithful to God. A graphic example of the predicted faulty reasoning is that we are coming to accept that it is perfectly normal to use the anus (ironically, the source of feces) as a sexual outlet. This kind of thinking where the obvious is ignored is what the Bible calls "depraved reasoning" and is predicted as a byproduct of loss in faith in God.

One other area the data speaks about is pornography. The internet has ushered in the ability for anyone to get access to anything their cravings desire. The cravings are not new: they have existed since time began. What is new is the ability to make this behavior "normal." The data shows that 40 million Americans regularly visit porn sites. Moreover, 35% of all internet downloads are related to pornography.

Again, the Bible teaches that we have the right to make these choices. However, we do not get to choose the consequences. In addition, the consequences are spelled out. Societies that turn from God ultimately collapse.

Since these trends are apparent in the US, the Bible predicts that our society would degrade in its ability to function. What does the data say about how well our society is working? In the interest of brevity, we choose the economic health of the nation as an indicator. Perhaps it is a coincidence, but in the same period as the decline in the religiosity index, births out of wedlock skyrocketed.

Acceptance of ultimate expressions of the meaning and purpose of sexuality skyrocketed. Moreover, our dependence on government has skyrocketed. We

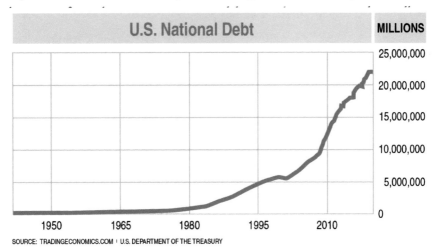

U.S. National Debt	MILLIONS

SOURCE: TRADINGECONOMICS.COM ¦ U.S. DEPARTMENT OF THE TREASURY

Perhaps all these happenings and the data are a coincidence. The Bible suggests not. It is predicted. The premise of the Bible is that God designed society to work in specific ways. In addition, He gave us the ability to choose alternatives. However, we do not get to choose the consequences. Please understand that the Bible makes clear that the problem is not debt, homosexuality, promiscuity, or pornography. The problem is a rejection of God Himself; the inability to overcome our weaknesses is just a symptom.

Prediction 4: Global Unprecedented Disasters

The Bible describes the "time of the end" as a series of events that build on each other and lead to the establishment of a "new heaven and earth"—the kind of peaceful world we dream of (prediction 7). One major predicted phase in this build up to the end is a period of major disasters. Some of them are a natural consequence of the actions of mankind and some are directly caused by God Himself. A major premise of the Bible is that nothing happens without God's knowledge and consent.

One of the items predicted that will cause major destabilization is the destruction of a prime economic power literally overnight ("in an hour" is the term used). This is such a big event that a whole chapter in the Bible is devoted to it. It is described as an economic power that consumes so much that the whole world sells to her. According to the Bible this country is predicted to be arrogant, wealthy, spoiled, and thinks she is "untouchable" (safe and secure).

God judges this nation for her arrogance and for spreading her arrogance to the world. God asserts his sovereignty by causing "destruction by fire" in a single hour. (As described in the war chapter, we now have the technology to destroy whole nations in an hour.) Also predicted is that the "merchants of the earth will weep and mourn" because of the loss of business. The sudden loss of this nation will have major impact on the global economy and stability of balance of power and political turmoil. [Note: This type of judgement is not unprecedented in the Bible. One example is the destruction of Sodom and Gomorrah recorded in a single day by fire and brimstone. Archeologists have found evidence of that civilization and its destruction by intense heat. One theory is that a meteorite exploded over the area.]

A separate disaster prediction is that something like a large meteorite ("a mountain from the sky") will hit the earth and cause such devastation to land and sea that a third of the life on the earth will be lost. This was written two thousand years ago before we even knew about meteorites. Now look at the headlines:

"City killer" asteroid misses Earth — and scientists had no idea

JULY 29, 2019 / 9:20 PM / CBS NEWS

Scientists revealed an asteroid dubbed by some as a "city killer" came closer to the Earth than the moon this week. The Washington Post reported that scientists apparently had no idea it was coming.

Asteroid 2019 OK came hurtling toward Earth at a speed of nearly 15 miles a second, before flying past. According to, it was about 45,000 miles from Earth on Thursday.

"It would have hit with over 30 times the energy of the atomic blast at Hiroshima," astronomer professor Alan Duffy told the Sydney Morning Herald. Duffy called the zooming space rock a "city killer."

Again, the Bible predicted two thousand years ago what with today's science we know is likely to happen eventually. Scientists estimate that several dozen asteroids in the 6–12 m (20–39 ft.) size range fly by Earth at a distance closer than the moon every year, *but only a fraction of these are actually detected.*

The Bible also predicts that there will be many other disruptive events in the end time period, including earthquakes, volcanos, famine, and disease. This is predicted to lead ultimately to global economic collapse and hyperinflation. It is not just the US that is moving in an unsustainable economic direction, it is the rest of the world as well.

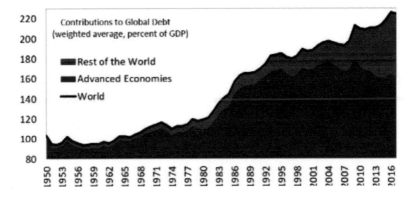

- Global debt has reached an all-time high of $184 trillion in nominal terms, the equivalent of 225 percent of GDP in 2017. On average, the world's debt now exceeds $86,000 in per capita terms, which is more than 2½ times the average income per-capita.

- The most indebted economies in the world are also the richer ones. You can explore this more in the interactive chart below. The top three borrowers in the world—the United States, China, and Japan—account for more than half of global debt, exceeding their share of global output.

Economies (and money itself) are based on trust. Trust that a piece of paper or entry in your bank account represents real value that can be exchanged later. The Bible predicted thousands of years ago that we would have more money than value. Hyperinflation represents the correction of this deception. This is another example of the principle described earlier. Governments can choose to inflate the money supply, but they cannot control the long-term consequences of doing so.

Prediction 5: Emergence of a One-world Government

The Bible predicted thousands of years ago that in the "time of the end," a one-world government would emerge. This seems like a highly unlikely event. Can you imagine China, Russia, or the US willingly giving up its sovereignty? Something must happen to make this scenario be acceptable to a very nationalistic world.

The Bible gives the answer. The predictions in the previous section describe a series of events so horrific that half the world's population is killed, economies die, wars, turmoil, and chaos reigns. This will make people willing to do anything to be fed and to have order. The Bible predicts that the outcome of this chaos will be a one-world government headed by a single individual. This government will put in place a new economic system that will be cashless. It will require people to buy and sell based on some sort of implant (or mark) in or on everyone's body.

Although predicted two thousand years ago, the technology is now in place to do this. The combination of the internet, RFID chips, and global banking networks makes this possible now. For this to happen cash must be eliminated as a legal exchange. Societies are rapidly moving in this direction willingly due to the convenience of online transactions. Canada leads the world today with more than 50% of its transactions being cashless. Fiji has announced an initiative to transform its whole society to be cashless. How many of us are taking advantage of Apple Pay or at least using credit cards for many purchases?

There are many good reasons to do this. Cashless societies lower cost, reduce theft, control illegal activities like the drug trade and provide just plain old convenience. How much of your shopping do you now do online? Who wants to go the bank anymore? It's so easy.

However, wherever there is convenience there is dependency. It is unavoidable. In the past 100 years, we have become completely dependent on electrical power, cars, gasoline, grocery stores, etc. These are convenient, but we no longer have the ability to live without them. For that matter, who would want to?

So it will be with a cashless society. It will be convenient, but the Bible predicts it will be an irresistible temptation for government to use this system to control people, especially those who dissent from the new world order. The Bible predicts that Christians will be targets of this control. Christians hold no higher allegiance than to Jesus Christ, which is a threat to the new world order. People will have to choose between following their God or this new world order that tries to eliminate different religions that are a source of disagreement, wars, and unrest. Global peace sounds like a worthy goal. However, unless that leader is God Himself, it will ultimately fail.

Prediction 6: The End of Tolerance

The American Heritage Dictionary defines tolerance as "the capacity for or the practice of recognizing and respecting the beliefs or practices of others." There is much discussion about what tolerance means and what its value is in society. For many people, tolerance has become like a system of faith. For them, tolerance is something to be valued in and of itself and valued above God Himself. The Bible predicts that God will eventually no longer tolerate false tolerance.

The Bible teaches to love your neighbor as yourself, which some think of as a mandate for tolerance. However, the Bible also clearly teaches that tolerating anything that violates God's design is deceiving. In other words, loving tolerance so much that embracing things that are not true is a deception and will ultimately fail. A simple example in the engineering world would be tolerating a different definition of what two plus two means. Good engineering is built on truth and so are societies.

What the Bible reveals is that for a period in human history God will tolerate the choices of mankind not to believe Him and ignoring His intended design. However, eventually this tolerance will end. There will come a time when there is a separating of those who have a relationship with God from those who do not.

Jesus is very specific and narrow on this topic. He says He is the *only* way to a relationship with God. According to modern society, this is a very "intolerant" view. C.S. Lewis summed this up when reviewing the claims Jesus made by saying you have to either categorize Jesus as a "liar, lunatic, or the Lord" of the universe. The choice is ours to make. There are other religions that make similar claims, that their way is the only way, which gives mankind even more choices.

The Bible predicts that all this confusion will be resolved. One day Christ will return and prove to all that He exists and that He has all power and authority over the universe. There will be no more "tolerance" of those who believe otherwise and have chosen a different path.

The Bible predicts that some people will react to this truth with a change of heart and entry into a relationship with Jesus Christ. In addition, others will be furious that what they want to be true is not. Finally, all differences between deceived and distorted thinking and behavior will be ended and the consequences for those beliefs will be understood.

The Bible also predicts that until Christ comes there will never be true peace in the world. When some people believe the truth and some people do not, there can never be long-term peaceful co-existence. Beliefs leads to actions and actions lead to long-term consequences. When Christ returns, He will resolve all this conflict and usher in the next prediction.

Prediction 7: Heaven on Earth

Last of all, the Bible predicts that eventually, there will be a new heaven and new earth. Moreover, in this place God will live amid those who have chosen a relationship with Him. Those who have rejected Jesus will not be in this place. There will be no more skepticism or unbelief. People will understand that God is their creator and the total source of their being and enjoy an unspeakable love and fellowship.

There will be no more death, tears, mourning, or pain in this place. The beauty of this place is beyond description. The Bible predicts construction where streets are made of pure gold and fine jewels. It speaks of massive size and wonder.

It is a place where there is no more darkness. The light comes from God Himself and is constant. Everyone in heaven will be truthful and honorable with no more fighting or hatred. In the presence of God, we will all understand our humble dependence upon Him, and comparisons and pride will be gone.

Most importantly, our identity will finally be understood and purified. All our failures to live the way God wanted us to—our jealousies, lies, hatred, yielding to wrong desires—will be removed. We will be stripped of all that is wrong with us and we will finally see the fulfillment of who we were created to be. We will even be given a new name by Christ Himself. Naming in the Bible is the privilege of a parent; it is a sign of ownership. Those who have children understand what it means to say, "this is my daughter" or "this is my son." It is the fulfillment of the deepest of relationships and an eternal connection that can never be broken.

The very last words of the Bible record Jesus' final message. He says, "Come. Let the one who hears come. And the let the one who is thirsty come; let the one who wishes take the water of life without cost."

Therefore, the Bible ends restating the final choice. Take the water of life or not. The choice is yours. (But not the consequences.)

CHAPTER TWELVE

Changing Times – The Millennial Generation

The future is molded by the past, and sometimes, age. The habits of today's millennials—those born between 1981 and 1996—may tell us what the future holds for this age group that will be between 54 and 69 in 2050. Today they represent over 22% of the U.S. population and range in age from 23 to 38. The age population distribution is reflected in the graph below. [91]

Population distribution in the United States in 2018,

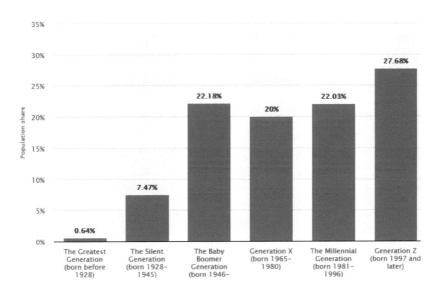

The millennials are tech savvy and fairly well-educated, which will be a plus at the mid-point of the 21st century. Retirement savings, on the other hand, will be less than adequate to support their past lifestyle. A recent article contained a list of several items millennials refused to spend their money on. [92] Cereal was on the list because it took too much time to clean up after eating it. Moreover, milk was on the list. Sorry kids, no coco puffs for breakfast.

Let's examine the list for more impactful items, followed by your author's own predictions fast forwarded to 2050.: [93]

- Movie Theaters
- Cars
- Cable TV
- Department Stores
- Life Insurance
- Houses
- The Stock Market
- Children

Movie Theaters

No surprise here. Go to a movie theater today and it looks like Senior Day. Millennials get their movie fix from Netflix, Amazon Prime, and Hulu. It's much easier to stay at home and watch your favorite movie, and much cheaper. Moreover, you don't have to take out a pay day loan to buy some popcorn.

This trend will continue in 2050. TVs will take up an entire wall and will provide a theater-like atmosphere in the home. Robbie the robot will serve the popcorn.

Cars

Cars are expensive to lease, own, and operate. Having a car competes with money needed for student debt repayment. Uber is a cheap alternative as is public transportation where available. Texting will summon your car, which will be driverless in a few years.

In 2050, car ownership in urban areas will be the exception rather than the rule. The Ubers of tomorrow will fly to your apartment building in a driverless car, pick you up, and take you to your destination.

Cable TV

Millennials are ditching cable TV for shows they can watch on Amazon Prime and Netflix.

TVs in 2050 will be truly "smart." They will entertain you and monitor your home and health. Your TV will know your preferences and deliver content to suite your tastes. Remotes will be gone and replaced by voice technology, a sophisticated version of today's Alexa.

Department Stores

Sears is on its way to being a historical footnote. Malls are closing or transitioning to boutique stores. Amazon and Wayfair are today's department stores. Millennials shop on their phones and tablets. And it's just not the younger generations that are bypassing the traditional department stores. All age groups are changing their shopping habits.

In 2050, department stores as we know them today will cease to exist. Virtual Reality will allow consumers to shop at home, try on clothes, and smell perfumes while sipping a glass of wine. Experts writing in The Future of Shopping report talk about the impact of the "fourth industrial revolution"—a merging of physical, digital and biological technologies—on shopping. [94]

Life Insurance

Data shows that only about 25% of millennials have life insurance. They cannot afford it.

This impacts on the future from several perspectives. First, life is not infallible. In a family situation, children may lose a parent and have no visible means of support other than Social Security. Secondly, life insurance companies use premiums to lend to those who build our buildings and infrastructure. Without this source of capital, alternatives must be found, usually at higher costs.

Life insurance in 2050 will be technology driven. The insurance agent of today will be replaced by icons on a phone, tablet, or TV screen which will explain, price, and electronically bill premium payments. New types of insurance will be available, including long term care insurance (LTC) with built in cost and availability safeguards. LTC insurance will be mandated and paid for by employer/employee contributions.

Houses

Millennials prefer mobility and owning a house is not the best mobility strategy. So, you either move in with your parents or rent an apartment. Also, young professionals are moving into urban downtown areas where entertainment, bars, coffee houses, and shopping is close by. This compounds the housing problems because disposable income is spent on having fun rather than paying off a mortgage.

With driverless and flying cars in 2050, living in suburbia will become more common as the commute from work will be relaxing. Homes will have rooftop garages covered with retractable domes. Inside the home, everything from appliances to security will be electronically controlled. Home ownership will become more attractive. Older homes will be occupied by the less affluent.

Flying cars will cause a dilemma for millennials. Should they live downtown where the action is, or in suburbia? Married couples will opt for the latter choice.

The Stock Market

How do millennials invest? Only 13% say they would use the stock market. Their choice? Cash, real estate, or gold. Either ultra conservative or risky are their investment goals. Hopefully, as they get older, they'll seek a more balanced investment strategy. If not, their retirement nest egg will be in jeopardy. According to a recent study, 66% of millennials have saved nothing for retirement. [95] Employers matching 401(k) or 403(b) programs are being ignored. This is sad. They need to purchase Suzi Orman, the financial planning guru's book.

According to Jeremy Slegel, in his book, *Stocks for the Long Run* (4th Edition), China will be the wealthiest nation in 2050. The world's stock markets are reflected in the chart below.

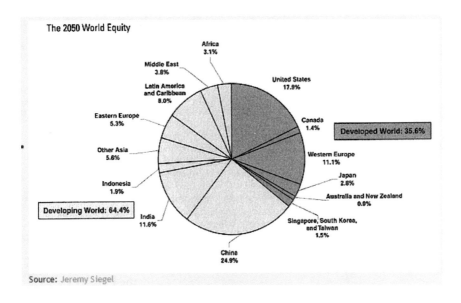

The 2050 World Equity

Source: Jeremy Siegel

Your authors are not about to predict the value of the stock market indices in 2050. Suffice it to say, today's pandemonium Wall Street traders will be gone from the exchange floor and trading will be done electronically. The image below will be relegated to the history books.

This Photo by Unknown Author is licensed under CC BY

Children

Hospital nurseries should not depend on millennials for their existence. Half of the millennial generation does not plan on having children, and those that do will have them later in life.

In the 2010 census, the percentage of children in the U.S. was at an all-time low of 24%. The percentage is expected to drop to 20.1% by 2049. [96] It is probable that the government in 2050 will have programs that encourage women to have children. Such inducements may include free day care or more favorable tax incentives.

Certainly, other generations will impact the future of the world and the U.S. The snapshot of the millennial generation was chosen as they will be the baby boomers of 2050.

ENDNOTES

1   Nostradamus Predictions. Quora. https://www.quora.com/topic/Nostradamus-Predictions

2   Predictions of the Future. October 32. 2018. https://www.24hourcampfire.com/ubbthreads/ubbthreads.php/topics/13245324/1/a-look-into-the-future

3   Levitt, K et al. National Health Care Expenditures, 1990. *Medicare and Medicaid Research Review.* https://www.ncbi.nlm.nih.gov/pmc/articles/PMC4193227/

4   Table 1. Sawyer, B. and Cox, C. How does health spending in the U.S. compare to other countries? *Kaiser Family Foundation.* (December 7, 2018). https://www.healthsystemtracker.org/chart-collection/health-spending-u-s-compare-countries/#item-start

5   US medical health research spending on the rise, but for how long? *Research America.* (November 15, 2017). https://www.researchamerica.org/news-events/news/us-medical-health-research-spending-rise-how-long

6   Top ten medical technology companies worldwide based on revenue in 2017 (in billion U.S. dollars). Statista. https://www.statista.com/statistics/281544/revenue-of-global-top-medical-technology-companies/

7   Paavola, A. 12 U>S. Cities ranked by cost per square foot to build a hospital. *Becker's Hospital Review.* (July 21, 2017). https://www.beckershospitalreview.com/facilities-management/12-us-cities-ranked-by-cost-per-square-foot-to-build-a-hospital.html

8   Vedinak, J. Skilled nursing construction costs continue to rise. *Skilled Nursing News.*

9   Raudaschi, A. Quantum Computing and health care. *BMJ Technology Blog.* (November 3, 2017). https://blogs.bmj.com/technology/2017/11/03/quantum-computing-and-health-care/

10 https://www.cbsnews.com/pictures/the-evolution-of-telephones/29/

11 https://www.apple.com/shop/buy-iphone/iphone-xr

12 https://upload.wikimedia.org/wikipedia/commons/d/d8/Macintosh_classic.jpg

13 The ultimate guide to understanding Augmented Reality (AR) technology. *Reality.* https://www.realitytechnologies.com/augmented-reality/

14 https://www.amazon.com/s?k=hololens&hvadid=77996658371707&hvbmt=b e&hvdev=c&hvqmt=e&tag=mh0b-20&ref=pd_sl_3dnuxfvhj_e

15 Hayes, E. Health Care of the future: Experts envision medicine in 2050. *Portland Business Journal.* (September 17, 2015). https://www.bizjournals.com/portland/blog/health-care-inc/2015/09/health-care-of-the-future-experts-envision.html

16 Genomics. *Merriam - Webster.* https://www.merriam-webster.com/dictionary/genomics

17 Hayes, E. ibid at https://www.bizjournals.com/portland/blog/health-care-inc/2015/09/health-care-of-the-future-experts-envision.html

18 The history of the pharmaceutical industry. *Pharmaceutical Chillers.* http://www.pharmaceuticalchiller.com/history-pharmaceutical-industry/

19 The Historic Drugstore. *Soderlund Drugstore Museum.* https://www.drugstoremuseum.com/drugstore/

20 He's the size of a golden retriever puppy and covered with scales. *National Geographic.* (June 2019). Pgs. 84-101

21 Harris, R. U.S. funding of health research stalls as other nations rev up. *NPR.* https://www.npr.org/sections/health-shots/2015/01/13/376801357/u-s-funding-of-health-research-stalls-as-other-nations-rev-up

22 Swanson, A. Big pharmaceutical companies are spending far more on marketing than on research. *The Washington Post.* (February 11, 2015). https://www.washingtonpost.com/news/wonk/wp/2015/02/11/big-pharmaceutical-companies-are-spending-far-more-on-marketing-than-research/?utm_term=.ecbe13054d91

23 Number of all hospitals in the U.S. from 1975 to 2017. *Statista.* https://www.statista.com/statistics/185843/number-of-all-hospitals-in-the-us-since-2001/

24 Share of old age population (65 years and over) in the total U.S. population from 1950 to 2050. *Statista.* https://www.statista.com/statistics/457822/share-of-old-age-population-in-the-total-us-population/

25 Wall, M. The disruptors. On the move. *BBC.* www.bbc.com/future/bespoke/the-disruptors/on-the-move/

26　CDC: 53 million adults in the U.S. live with a disability. *Centers for Disease Control and Prevention.* (July 30, 2015) https://www.cdc.gov/media/releases/2015/p0730-us-disability.html.

27　Baggely, K. 6 surpriing ways driverless cars will change our world. Are you ready? *NBC News.* (April 18, 2018). https://www.nbcnews.com/mach/science/6-surprising-ways-driverless-cars-will-change-our-world-ncna867061

28　Michelsen, C. Austria mates parking meters to EV charging stations. *Gas 2.* (September 11, 2011) https://gas2.org/2011/09/14/austria-mates-parking-meters-to-ev-charging-stations/

29　https://www.nationalgeographic.com/environment/2019/06/midwest-rain-climate-change-wrecking-corn-soy-crops/

30　World population expected to reach 9.8 billion in 2050, and 11.2 billion in 2100. *United Nations.* (June 21, 2017). https://www.un.org/development/desa/en/news/population/world-population-prospects-2017.html

31　Babones, S. China's high-speed trains are taking on more passengers in Chinese New Year massive migration. *Forbes.* (February 13, 2018). https://www.forbes.com/sites/salvatorebabones/2018/02/13/chinas-high-speed-trains-are-taking-on-more-passengers-in-chinese-new-year-massive-migration/#79d81b7a423f

32　*The Boring Company.* https://www.boringcompany.com/products

33　https://news.sky.com/story/blue-origin-amazons-jeff-bezos-unveils-plans-to-send-a-spaceship-to-the-moon-11715867

34　Hincks, J. The world is headed for a food security crisis. Here's how we can avert it. *Time.* (March 28, 2018). https://time.com/5216532/global-food-security-richard-deverell/

35　Ibid

36　Hunter, M. We don't need to double world food production by 2050 – here's why. *The Conservation.* (March 8, 2017). http://theconversation.com/we-dont-need-to-double-world-food-production-by-2050-heres-why-74211

37　Genetically modified foods. *Learn. Genetics.* https://learn.genetics.utah.edu/content/science/gmfoods/

38　Ibid

39　A comparison of world corn yields. *AGPRO.*

40　Impossible Foods https://impossiblefoods.com

41   visions of the future: Life on Mars and a rural skyscraper. *MACH*. (May 26, 2019). https://www.nbcnews.com/mach/news/visions-future-life-mars-rural-skyscraper-ncsl1004486

42   http://discoversinai.net/english/bedouin-tent-in-the-past-and-today/8736

43   -year period that changed cities forever – The Veterans' emergency housing program. *City Clock*. (June 17, 2014). http://www.cityclock.org/program-changed-cities-forever/#.XQu0X25FyUk

44   https://www.amazon.com/Allwood-Eagle-Point-1108-Cabin/dp/B00LYGIEU2/ref=sr_1_7?hvadid=78065425152567&hvbmt=be&hvdev=c&hvqmt=e&keywords=tiny+house&qid=1561050319&s=gateway&sr=8-7

45   visions of the future: Life on Mars and a rural skyscraper. *MACH*. (May 26, 2019). https://www.nbcnews.com/mach/news/visions-future-life-mars-rural-skyscraper-ncsl1004486

46   Peach, J. What will a typical 2050's home be like? *Smart Cities Dive*. https://www.smartcitiesdive.com/ex/sustainablecitiescollective/what-will-typical-2050s-home-be/22109/

47   Futurology. The new home in 2050. *NHBC Foundation*. https://www.nhbcfoundation.org/wp-content/uploads/2018/05/NF80_Futurology-1.pdf

48   visions of the future: Life on Mars and a rural skyscraper. *MACH*. (May 26, 2019). https://www.nbcnews.com/mach/news/visions-future-life-mars-rural-skyscraper-ncsl1004486

49   Ibid

50   Ibid

51   Andreea, C. Top 27 future concepts and gadgets for the home of 2050. *Homesthetics*. (December 24, 2014). https://homesthetics.net/top-27-future-concepts-and-gadgets-for-the-home-of-2050/

52   Durden, T. "It's a perfect storm": List of retailers in danger of bankruptcy hits record 22. *Zero Hedge*.(June 9, 2017). https://www.zerohedge.com/news/2017-06-09/its-perfect-storm-list-retailers-danger-bankruptcy-hits-record-22

53   Liu, G. The best food delivery apps of 2019. *Digital Trends*. (March 11, 2018). https://www.digitaltrends.com/home/best-food-delivery-apps/

54   Soper, S. Amazon will consider opening up to 3,000 cashierless stores by 2021. *Bloomberg*. (September 19, 2018). https://www.bloomberg.com/

news/articles/2018-09-19/amazon-is-said-to-plan-up-to-3-000-cashierless-stores-by-2021

55 Kidd, T. Online retail in 2050. *Kantar.* (March 12, 2012). https://uk.kantar.com/business/retail/online-retail-in-2050/

56 Von Radowitz, J. Virtual reality to replace High Street shopping by 2050. *Independent.* (November 28, 2016). https://www.independent.co.uk/news/business/news/virtual-reality-to-replace-high-street-shopping-by-2050-a7442221.html

57 Muoio, D. 17 ways technology will change our lives by 2050. *Business Insider.* (July 8, 2016). https://www.businessinsider.com/ian-pearson-predictions-about-the-world-in-2050-2016-7

58 How do 3D printers work? *Department of Energy.* (June 19, 2014). https://www.energy.gov/articles/how-3d-printers-work

59 Work. *Merriam-Webster.* https://www.merriam-webster.com/dictionary/work

60 Andriotis, N. The 2020 workplace – the future workplace trends you should know right now. *Talent Lms.* (July 31, 2017). https://www.talentlms.com/blog/2020-workplace-trends-hr/

61 Moore, A. A complete guide to Apple Park. *Macworld.* (February 20, 2018). https://www.macworld.co.uk/feature/apple/complete-guide-apple-park-3489704/

62 jobs that will disappear in the next 20 years due to AI. *Alux.com.* (October 5, 2017). https://www.youtube.com/watch?v=r211u89eUaY

63 Perry, P. 47% of jobs will vanish in the next 25 years, say Oxford University researchers. *Big Think.* (December 24, 2016). https://bigthink.com/philip-perry/47-of-jobs-in-the-next-25-years-will-disappear-according-to-oxford-university

64 Ibid

65 Stanger, M. 9 ways the workplace will be different in 2050. *MSN/Money.* (January 21, 2016). https://www.msn.com/en-us/money/careersandeducation/9-ways-the-workplace-will-be-different-in-2050/ar-BBovdKg

66 Job Market in 2050. *Huff Post.* (December 6, 2017). https://www.huffpost.com/entry/job-market-in-2050_b_7516968

67   Novoseltseva, E. Tech of the future: technology predictions for our world in 2050. *Apiumhub* (January 9, 2018). https://apiumhub.com/tech-blog-barcelona/tech-of-the-future-technology-predictions/

68   What will classroom teaching look like in 2050. *Toppr Bytes*. (September 11, 2016). https://www.toppr.com/bytes/classroom-teaching-in-2050/

69   Harper, A. Will robots replace teachers in the future? *EducationDive* (November 15, 2018). https://www.educationdive.com/news/will-robots-replace-teachers-in-the-future/542239/

70   Bitcoin. *Wikipedia. https://en.wikipedia.org/wiki/Bitcoin*

71   Marley, R. 10 problems with cryptocurrency that prevent its mainstream use. *Chain*. https://thechain.media/10-problems-with-cryptocurrency-that-prevent-its-mainstream-use

72   List of countries in the European Union. *World Population Review*. http://worldpopulationreview.com/european-union-countries/

73   United States one hundred-dollar bill. *Wikipedia*. https://en.wikipedia.org/wiki/United_States_one_hundred-dollar_bill

74   Iskyan, K. This is the real reason China is buying gold (hint: it's not so exciting). *Stansberry Pacific*. (May 4, 2016). https://stansberrypacific.com/china/this-is-the-real-reason-china-is-buying-gold-hint-its-not-so-exciting/

75   Kagan, J. Financial technology - Fintech. *Investopedia*. (June 25, 2019). https://www.investopedia.com/terms/f/fintech.asp

76   Ibid.

77   The world in 2050. *PWC*. https://www.pwc.com/gx/en/issues/economy/the-world-in-2050.html#downloads

78   ibid

79   Causes of global warming, explained. *National Geographic*. (January 17, 2019). https://www.nationalgeographic.com/environment/global-warming/global-warming-causes/

80   Global Climate Change. *NASA*. https://climate.nasa.gov/vital-signs/carbon-dioxide/

81   Harris, C. and Mann, R. Global temperature trends from 2500 B.C. to 2040 A.D. *Long Range Weather*. (March 10, 2018). http://www.longrangeweather.com/global_temperatures.htm

82   The history of the Ford River Rouge plant. *The Old Motor*. (January 31, 2013). http://theoldmotor.com/?p=64735

83   Brixey-Williams, S. Which countries do the most to protect the environment? *World Economic Forum*.(September 30, 2015). https://www.weforum.org/agenda/2015/09/which-countries-do-the-most-to-protect-the-environment

84   Santiago, J. These countries are best prepared for climate change. *World Economic Forum*. (August 25, 2015). https://www.weforum.org/agenda/2015/08/these-countries-are-best-prepared-for-climate-change/

85   Pascus, B. Human civilization faces "existential risk" by 2050 according to new Australian climate change report. *CBS News*. (June 4, 2019). https://www.cbsnews.com/news/new-climate-change-report-human-civilization-at-risk-extinction-by-2050-new-australian-climate/

86   Gilpin, L. 10 ways technology is fighting climate change. *TechRepublic*. (August 6, 2014). https://www.techrepublic.com/article/10-ways-technology-is-fighting-climate-change/

87   Gray, A. 5 tech innovations that could save us from climate change. *World Economic Forum*. (January 9, 2017). https://www.weforum.org/agenda/2017/01/tech-innovations-save-us-from-climate-change

88   https://en.wikipedia.org/wiki/War

89   https://en.wikipedia.org/wiki/War

90   https://arstechnica.com/information-technology/2017/12/dubious-claim-of-week-air-forces-emp-missile-could-disable-n-korean-icbms/

91   Population distribution in the United States in 2018, by generation. *Statista*. https://www.statista.com/statistics/296974/us-population-share-by-generation/

92   Katie, B. These items are vanishing because millennials refuse to buy them. *ARTICLES VALLY*. (JULY 24, 2019). HTTP://ARTICLESVALLY-PROD-DOCKER.RDP2NTTWKD.US-WEST-2.ELASTICBEANSTALK.COM/WORLDWIDE/MILLENNIAL

93   Ibid

94   Von Radowitz, J. Virtual Reality' to replace high street shopping by 2050. *Independent*. (November 28, 2016). https://www.independent.co.uk/news/business/news/virtual-reality-to-replace-high-street-shopping-by-2050-a7442221.html

95   Lobosco, K. 66% of millennials have nothing saved for retirement. *CNN Money.* (March 7, 2018). https://money.cnn.com/2018/03/07/retirement/millennial-retirement-savings/index.html

96   Children as a percentage of population. *ChildStats.gov.* https://www.childstats.gov/americaschildren/tables/pop2.asp